Coal—A Window to Past Climate and Vegetation

Miryam Glikson-Simpson

Coal—A Window to Past Climate and Vegetation

Springer

Miryam Glikson-Simpson
Palmwoods, QLD, Australia

ISBN 978-3-030-44474-7 ISBN 978-3-030-44472-3 (eBook)
https://doi.org/10.1007/978-3-030-44472-3

Cover illustration: Ludek Pesek Science Photo Library

This Springer imprint is published by the registered company Springer Nature Switzerland AG
The registered company address is: Gewerbestrasse 11, 6330 Cham, Switzerland

In honor of ***Prof. G. H. Taylor*** *Founder of* ***modern coal petrology****. Supervisor of my Ph.D. thesis, 1982; then Director of the Centre for Resource and Environmental Studies at the* ***Australian National University***

Acknowledgements

The samples for the pollen study were collected from the Geological Survey of Western Australia's core shed in Collie, and assistance and advice of Prof. B. E. Balme is gratefully acknowledged. The University of Western Australia provided laboratories and research facilities that enabled me to carry out this study.

Study and research of Spontaneous Combustion were carried out at the University of Queensland in collaboration with the Department of Mining Engineering during my time on the staff of the Department of Earth Sciences.

Many thanks to Dr. Judy Owen for palynological discussions, review and constructive comments.

Thank you to Ann Ferguson and Omair Raza for valued technical assistance.

Dr. Petra van Steenbergen's encouragement is much appreciated.

Special thanks are due to Emeritus Professor Rod Simpson for continuous encouragement and advice on many technical aspects arising throughout the laborious writing of this book.

Many thanks to Senior Editor Mrs. Margaret Deignan and Dr. Madanagopal Deenadayalan who made this work a reality.

Introduction

Gondwana, which includes Australia, South Africa, India, Antarctica and South America, holds the largest coal deposits. These coal deposits were formed from vegetation growing in the vast swamps that occupied huge slowly subsiding depressions left bare after the retreating glaciers at the end of the carboniferous. The early Permian climate was cold, with freezing winters and humid cold summers. Cold temperate climate followed for some time until the warming up towards the end of the Permian. The vegetation that gave rise to the vast coal deposits (Fig. 1) was predominantly the Glossopteris flora which was composed of trees varying in size and shedding their leaves in winter. The trunks of these plants and large quantities of their leaves settled in the marshes and degraded into peat which accumulated, and through diagenetic processes, subsidence and compression formed the coal deposits. The Permian vegetation shed their pollen and spores as well, and these were deposited with the rest of organic remains. Pollen and spores are resistant to degradation and their fossil remains are used to reconstruct the environment and landscape at the time of deposition, the climatic conditions and type of vegetation. The pollen and spores were extracted from coals that were formed from plant remains deposited during the Permian era that signifies the last period of the Palaeozoic between 298 and 250 million years ago: The end of the Permian marks a dramatic change in the flora in the wake of a meteorite impact in Brazil and extensive flood volcanism in Siberia and China, which triggered the release of immense sulphur-di-oxide fumes and carbon-di-oxide, resulting in the greatest known mass extinction of fauna and certain flora.

Pollen and spores in this book are described according to a devised and universally accepted systematic code; a critical approach of the traditional systematic classification is undertaken. Furthermore, pollen and spores are discussed in terms of their botanical affinity and are compared with pollen and spores of living Australian plants. Some of the Permian pollen/spores show close resemblance to those of Australian plants growing today. These signify continuity, whereas many others became extinct over a period of time as a result of climatic changes occurring in their environment. Ignoring the affinity of fossil pollen/spores to existing plants on one hand, the attempt to reconstruct past environments is incompatible.

Fig. 1 4 m thick coal seam in the Bowen Basin, Queensland, Australia

The composition of plant assemblages and their changes over time reflect changes in the depositional environments of the coal-forming plants, which in turn follow the climatic changes. These are a subject of this book. Comparison to present day environments and climates is made possible by identifying the remains of ancient flora.

Over 200 samples from bore holes in the Perth Basin and the Collie–Wilga Basins of Western Australia ranging from the Lower Permian to Upper Permian were processed for pollen analysis. These yielded groups of pollen and spores that represent the vegetation of the time.

Assemblages as well as individual species can also be used as markers signifying specific stratigraphic horizons in time and space.

Over 70 species of pollen and spores are described in detail, some of those described in earlier studies are compared with the ones in the present study, their affinity discussed and some new forms of pollen previously unknown are also described.

Contents

Chapter 1
General Review of Permian Sediments in Western Australia

Permian sediments are known from all the main sedimentary basins in Western Australia, including the Carnarvon, Canning, Bonaparte Gulf and Perth Basins, where Permian sequences reach more than 5000 m in thickness (McWhae et al. 1958). Permian rocks are exposed in the northern part of the Perth Basin, where their thickness approximates 2000 m (McWhae et al. 1958) and in the Collie and Wilga Basins south-east of Perth. Lower Permian glacial sediments outcrop in the Officer Basin (Interior Plateau) north of the Eucla Basin and south-east of the Canning Basin.

Sediments of Permian age were encountered in a number of bores in Western Australia; among them were Nullarbor No. 8 bore in the Eucla Basin (Harris and Ludbrook 1966), Sue No. 1 bore and Alexandra Bridge bore both in the Donnybrook Graben. The latter are dealt with in this book.

With the exception of the Eucla Basin, the Permian successions in Western Australia start everywhere with glacial sediments, namely tillites, deposited by the retreating glaciers. The character of the sediments overlying the tillites varies in different areas. Generally, the facies become progressively Marine towards the north, where the sediments reach their greatest thickness in the Carnarvon and Canning Basins. Shallow marine deposits occur in the northern Perth Basin and continental beds become important towards the southern and south-western parts of the Perth Basin. Only Lower Permian (Artinskian) sediments were encountered in Nullarbor No.8 bore in the Eucla Basin; this sequence is considered to be lagoonal with restricted connection to an open marine environment (Harris and Ludbrook 1966). A continental Permian sequence starting with tillites has been encountered in Barcoo-Junction No. 1 borehole in the Cooper Basin (Glikson; Unpublished Company Report 1980).

No marine fauna has been found in the Permian sediments of either the Collie–Wilga Basins or the bores in the southern Perth Basin. The Permian sequence of the Collie Basin closely resembles the Permian sequence in the southern Perth Basin in lithology as well as microflora, which suggests a common depositional basin for both. The Darling fault separating the two basins today became active sometime during the Permian: the Permian sequence in the Collie Basin (the uplifted block) on the eastern side of the fault being less than 1000 m thick compared with the close

M. Glikson-Simpson, *Coal—A Window to Past Climate and Vegetation*,
https://doi.org/10.1007/978-3-030-44472-3_1

to 6500 m thick Permian sediments in Sue No.1 bore in the South Perth Basin on the western (down-faulted) side of the Darling fault. The sediments on both sides closely resemble each other, representing the true span between the Sakmarian base of the Permian and the Upper Permian. On the whole, the continental Permian Basins of Collie–Wilga and south-western Perth Basin may have well constituted a south-western extension of a continental block located between the Eucla Basin in the south and the northern Perth Basin, the Canning Basin and Carnarvon Basins in the north and north-west, the so-called 'Interior Plateau' of McWhae et al. (1958). Based on palynostratigraphic correlations, Backhouse (1990) also suggests that the Collie and South Perth Basin coal measures were developed over much of the southern Yilgarn block and beyond. The northern Perth Basin on the other hand comprises interchanging and intergrading continental and marine deposits, thus representing a transition between the continental facies of the 'Interior Plateau' and the marine facies of the Carnarvon Basin.

Key Map

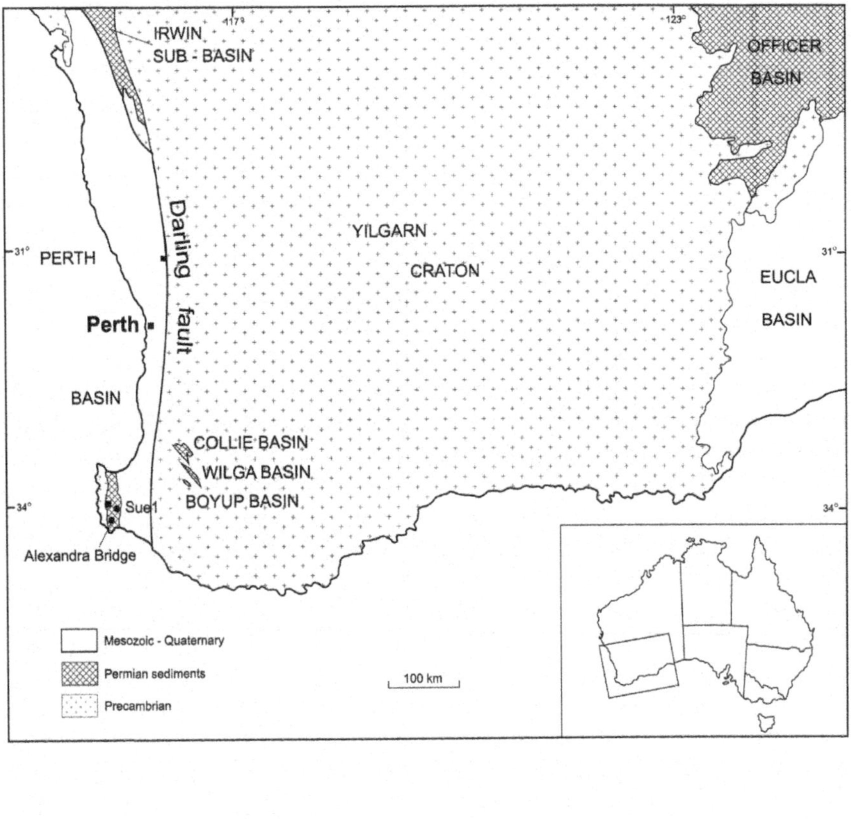

1.1 Previous Studies of Permian Pollen/Spores in the Collie and South Perth Basins

While a number of comprehensive studies and biostratigraphic correlations of Permian pollen in Australian Basins were published (Foster 1982; Price 1983; Mory and Backhouse 1997; Kemp et al. 1977), only a few studies were carried out on the pollen/spores of the Collie Basin. The earliest were pioneered by Balme (1952, 1959), Balme and Hennelly (1955, 1956a, b). More recent studies by Backhouse (1991; 1993) were detailed and covered both, the pollen and spore descriptions as well as stratigraphic correlations within Collie Basin and between Collie and Perth Basin. The present study focuses on the Collie Basin, including the smaller Wilga Basin next to it and South Perth Basin.

1.2 The Collie Basin

The Collie Basin in Western Australia (Key map, Fig. 1.1) is situated about 160 km south of Perth and about 22 km east of the Darling scarp. It is centred on the coordinates 33.5° S latitude, 116.4° E longitude. The basin forms a NW–SE-oriented depression in the Precambrian basement and is drained by the Collie River. The area rises from 200 m to about 250 m above the sea level and is about 22.5 km long and 12.8 km wide. The Collie Basin was considered a remnant of "…an erosional plain sloping gently to the Eucla Basin in the east, truncated by the Darling scarp in the west" (Hills 1961). This plain was assumed by Hills (1961) to be a low-lying land in the Permian. The origin of the depression was attributed to erosion by glaciers (Lord 1952). The detailed study of the Collie Basin by LeBlanc Smith (1993) describes the basin as a "fault bounded, post-depositional pull apart structure". The detailed study of LeBlanc Smith supports Lord's glacially eroded depression, filled with sediments, subject to later, post-depositional faulting. Similar conclusions of post-depositional faulting of a coal basin have been reached by Ghosh (2002) in his study of the Raniganj Basin in India, another Gondwana Permian post-glacial depositional basin.

The Collie Basin is 226 km^2 in area (LeBlanc Smith 1993) and is divided into two parts by a granitic-gneiss ridge (see map, Fig. 1.2): the eastern part of which is sub-divided again into a northern depression and a southern depression, termed the Shotts and the Muja, respectively. LeBlanc Smith (1993) states that a stratigraphic section of about 1400 m of faulted Permian sediments is preserved in the basin, of which over 900 m are coal-bearing.

Coal samples were examined from bores in the Collie Basin, sites A and B (text Fig. 1.2), and from the Muja sub-basin sites C and D and Wilga Basin Bore No. 1. The Permian sequence in the Collie Basin as well as the Perth Basin starts with *tillites* nonconformably overlying the glacially striated basement. These tillites are correlated on lithological and microfloral basis with the Nangetty Formation in the northern

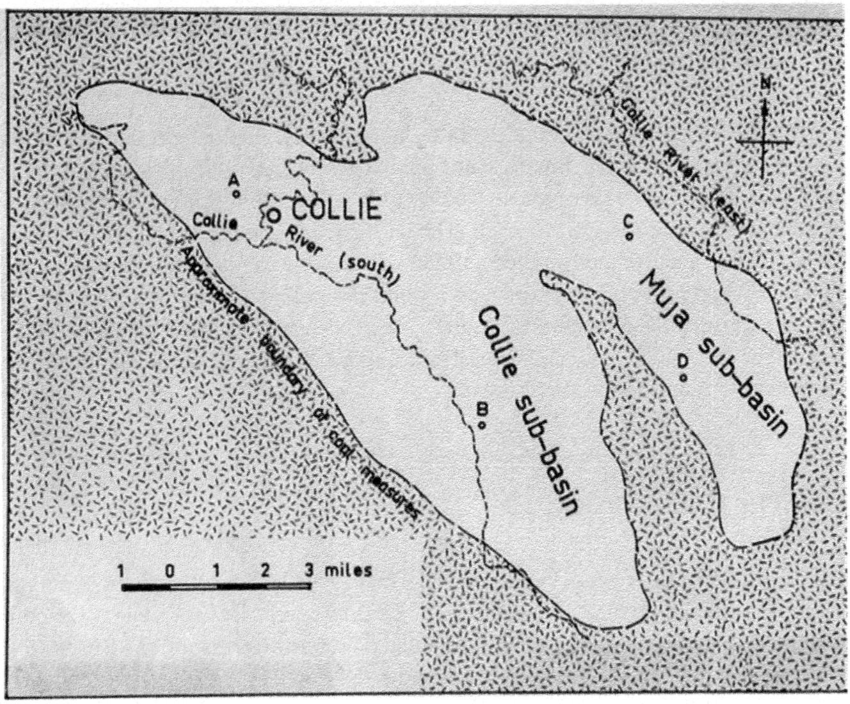

Fig. 1.1 Collie Basin

Perth Basin, the Lyons Group in the Carnarvon Basin and the Grant Formation in the Canning Basin, whose age is thought to be Sakmarian. Upper Carboniferous rocks were encountered below the Grant Formation in the Canning Basin (Balme 1959). The tillites in the Collie Basin are predominantly greenish grey mudstones and siltstones, referred to sometimes as the 'Stockton Formation' (Low 1958). The Stockton Formation varies in thickness from a few meters to about 212 m, being close to 100 m thick in Site B Bore of Collie and 10 m thick in Site D Bore of Muja. The formation is overlain by the Collie Coal Measures which are over 900 m thick in the Collie sub-basin and 67 m thick in the Muja sub-basin (LeBlanc Smith 1993). The Lower Collie Coal Measures have been correlated with the Irwin River Coal Measures, which are believed to be of Artinskian age (Balme in McWhae et al. 1958). The Irwin River Coal Measures overlie the high cliff sandstones, which have been dated as Upper Sakmarian (Glenister and Furnish 1961).

The upper or Collie Burn horizon is considered to be of Upper Permian age. The Permian strata in Collie Basin are overlain by Cretaceous deposits (LeBlanc Smith 1993).

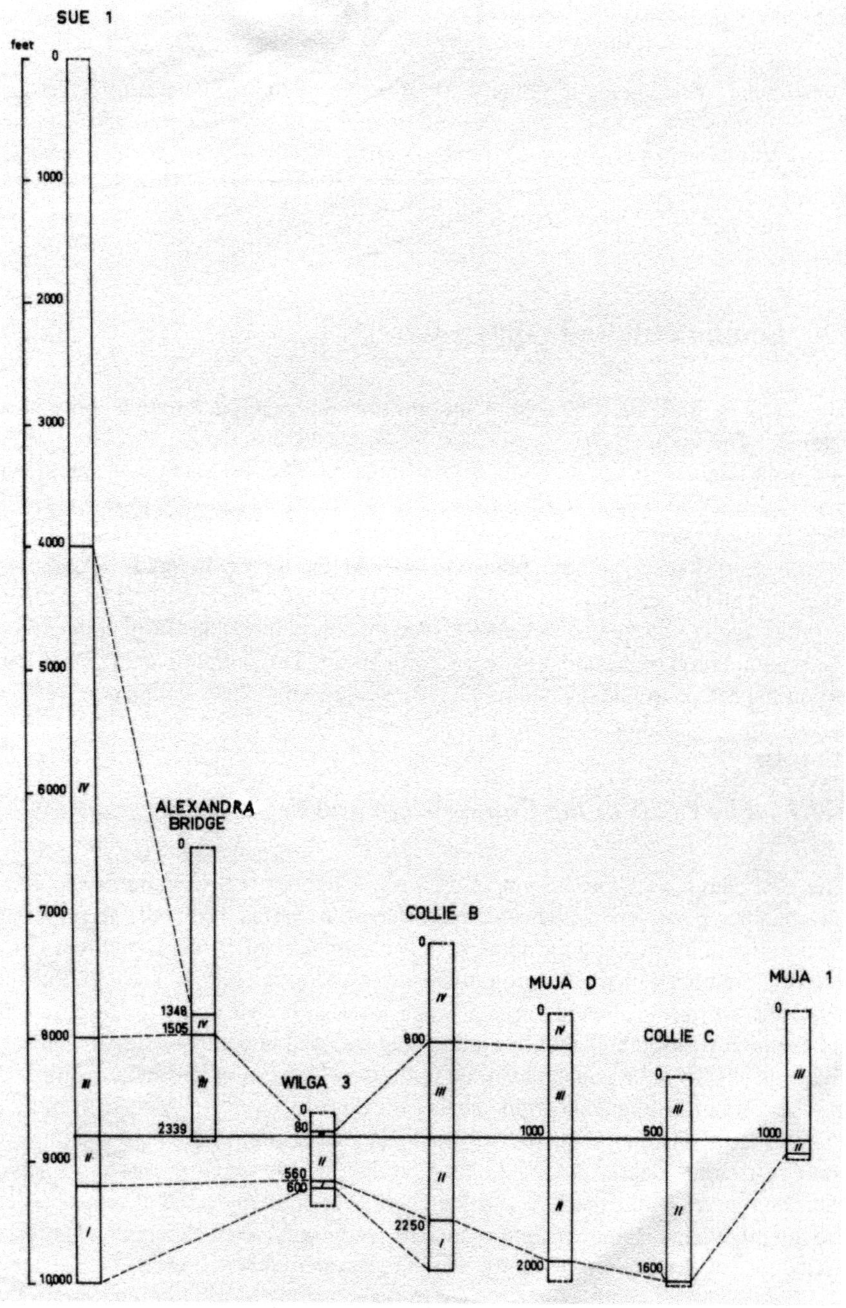

Fig. 1.2 Correlation diagram of boreholes in Collie and South Perth Basins

1.3 Wilga Basin

The smaller Wilga Basin is situated 40 km south of the Collie Basin (centred on 33° 6′ S latitude; 116.4° E longitude) and is smaller in size. The Permian sequence in the Wilga Basin starts with *tillites* at a depth of about 200 m, which are overlain by coal measures. The entire sequence contains plant microfossils of Sakmarian to Artinskian age. The Upper Permian is missing in the Wilga Basin.

1.4 South Perth Basin (Sue-1 Bore)

The South Perth Basin is marked by un-interrupted deposition throughout the entire Permian. The Permian coal measures in S.P.B. are 4653 m thick compared to about 900 m in Collie B bore and about 700 m in Collie D bore. Sakmarian samples in Sue-1 samples are 120 m thick compared to the 82 m thick tillite deposits in Collie B bore, and 12 m thick tillites in Collie C bore.

The Artinskian biostratigraphic units II and III are represented more or less equally in all Collie-Muja basin bores (Fig. 1.2). On the other hand, unit IV of Upper Permian age is represented in a diminished thickness in Collie Basin compared to the accumulation of around 2000 m in Perth Basin. This supports the Darling fault becoming active around the end of the Artinskian–early Upper Permian.

1.4.1 The Coals of the Collie–Wilga and Perth Basins

The coal samples of the present study were generally rich in exinites supplying the abundant pollen/spores; vitrinite/inertinite ratios varied. Generally, the coals are seen as inertinite-rich and vitrinite-poor. The samples usually displayed thin vitrite layers alternating with thick inertite/fusite layers. The high ratio of inertinite/fusinite relative to vitrinite in all Permian Gondwana coals has been traced to the climate at the time, which was wet and cool in summer and freezing in winter. Oxidation through freeze-drying of the organic matter led to the development of inertinite/fusinite.

The inertinite-abundant dull coals or *durains* were particularly rich in pollen/spores. The latter due to their lipid-dominated composition are significantly more resistant to degradation and oxidation than woody remains. Bright coals, vitrains dominated by the maceral vitrinite, were poor or often devoid of spores/pollen. On the other hand, some vitrinite-rich coals were found to be exceptionally high in acritarchs, supporting a water body, and higher water table at times.

The pollen and spore exines belong to the macerals of the ***Liptinite*** group that originates from H-rich lipidic (oily) plant material. The liptinites are relatively stable and therefore are resistant to biodegradation; hence they preserve in sediments, and are useful in reconstructing ancient climates and vegetation. *Liptinites* when observed in fluorescence mode give bright yellow glow at low maturity and become darker and duller as the coal increases in maturation, namely from sub-bituminous to bituminous, when the first distinct change in the lipid properties appears (Teichmuller 1974). This change coincides with petroleum formation and its discharge from *liptinite* macerals (Glikson and Fielding 1991), specifically from exines of pollen (Glikson and Owen 1987).

Usually migration of oil, which tends to be 'heavy' oil from coal, is limited and it tends to solidify as bitumen in cavities within *inertinite* macerals and coal cleats, as has been documented from the Permian coals of the Bowen Basin (Glikson et al. 1999, 2000).

The maturation indices used by LeBlanc Smith (1993) for the Collie Basin coals are vitrinite reflectance; reporting Ro% of 0.4–0.6 over a depth of about 1000 m. The higher values are reported from the base of the Permian coal in the Collie Basin. Taylor et al. (1998) reported VRo values of 1.0% from 4653 m depth, at the base of the Permian in the Perth Basin, the down-faulted block. The same authors note the basin as 'cool', namely having a low thermal gradient. LeBlanc Smith (1993) reported 21–31 °C geothermal gradient presently in the Perth Basin. Temperatures of just over 100 °C at the base of the Permian are reported by Taylor et al. (1998), which fits well with LeBlanc's geothermal gradient. The Perth Basin was part of the depositional basin that included Collie–Wilga; therefore, the same geothermal gradient applies.

Vitrinite reflectance data indicate that significant sections of Permian sediments are missing from the Collie Basin Coal Measures. Vitrinite Ro% with depth has been used to reconstruct the burial and tectonic history of sedimentary basins (Glikson and Golding 1998; Glikson et al. 2006). The jump in vitrinite Ro% from 0.4 to 0.6 (LeBlank Smith 1993) signifies missing sediments. It is therefore evident from the VRo data that the Darling fault became active initially sometime during the Late Artinskian–early Upper Permian. Hence the missing sediments in the uplifted block of the Collie Basin.

Vitrinite reflectance data reported by LeBlanc Smith (1993) for the coals did not display any significant different values between R_{min} and R_{max}, which indicates absence of anisotropy, the result of stress through lateral tectonic or hydrothermal pressure, which would have led to folding and distortion of the sediments (Glikson et al. 2006; 2011).

1.5 Methods of Study

Preparation of coal samples for pollen/spore analysis followed techniques outlined by C.A. Brown (1960) in 'Palynological Techniques'.

Preparation of Slides

Slides were mounted in glycerine jelly and sealed with 'Hemiglu'. Single mounts were prepared following Erdtman's (1934) method.

Microtome Sections

Selected specimens were sectioned following the method developed by Dettmann as outlined in: Hughes et al. (1962). Sectioning of fossil organic matter remains has been later updated, modified and applied by the author to ultra-sectioning of organic remains concentrated from coals and other sediments for observation in transmission electron microscopy (Glikson 1982; Glikson and Taylor 1986; Glikson 2001; Glikson 2008; Glikson et al. 2011).

1.5.1 Classification

The description of pollen/spores in this book constitutes an attempt to construct a simplified supra-generic classification for the accommodation of dispersed pollen and spores within sediments. Various different classifications have been used over time, with the main ones outlined and discussed by Dettmann (1963). For example, it is evident when viewing Potonie's (Potonie and Kremp 1954) and Bharadwaj's supra-generic classifications (Bharadwaj 1955, 1958, 1962) that the various groups, sub-groups and so on become more and more complex with the continual additions of new supra-generic divisions. The basis for these divisions and sub-divisions is usually the shape of the tetrad scar and/or *exine* sculpture. These criteria led to the splitting of form genera and even species and subsequently their distribution among different divisions and sub-divisions. Needless to say that this becomes impractical for correlation purposes which are one of the usages of fossil pollen and spores, and most impractical when attempting to interpret vegetational and climatic conditions.

It has to be taken into consideration that we are dealing with dispersed spores and pollen, and consequently lack adequate means for defining a genus in the sense of recent plant taxonomy. This fact leads to the possibility of fossil dispersed spores/pollen being grouped under different species of one genus which in fact could be just variations of one and the same species. This is demonstrated by studies of spores in situ; for example, a species of the fern *Senftenbergia plumosa* produced spores with bacculate projections (Radforth 1939; Remy 1954/6) resembling the dispersed Permian spore *Raistrickia*. On the other hand, another species of *Senftenbergia, S. pennaeformis* produced 'verrucate' spores similar to *Verrucosisporites pseudoreticulatus* and *Campotriletes biornatus*. This is an example of two species of one and the same genus in *sporae in situ* analogous to three genera and three species in *sporae dispersae*.

Smith (1962) in discussing various spores that are isolated and found within one sporangium emphasises the problem of *sporae dispersae* as follows: "As sporae dispersae, Type 'B' spores would probably have been placed in the genus *Convolutispora* Hoffmeister and Staplin (1960). Type 'A' spores do not appear to resemble any previously described spore type. Certainly the two types would have been widely separated in any classification. Both would most probably have been described under different names. It is likely that type 'A' would have been further split into two or more species".

The diversity in size and sculpture of spores found within one and the same sporangia was confirmed by the studies of Eggert and Taylor (1966): "The spores exhibit a remarkable diversity in size and ornamentation ranging from smooth-walled immature forms to mature, highly ornamented types". The variation within one species in Eggert and Taylor's study shows similarities to the fossil spores of *Raistrickia* and *Apiculatisporites* on one hand, and to *Punctatisporites* and *Calamospora* on the other hand.

Lindstrom et al. (1997) examined and studied pollen extracted from single complete Glossopterid sporangia and found an enormous variation in the pollen morphology.

The problem with the present intricate classification (e.g. Potonie) in use for fossil dispersed spores and pollen was discussed and summed up by Balbach (1966) as follows: "*Although current systems of plant microfossil classification are usable stratigraphically, the lack of a botanical taxonomic basis has led to the naming of a multiplicity of species, many of doubtful biological validity. There should be no reason why a botanically valid classification may not be made usable stratigraphically and vice versa*".

Indeed, recent studies of spores and pollen isolated from fossil sporangia and pollen sacs attached to fossil plants proved their affinity and connection to particular trees and shrubs, as well as displaying considerable variation in shape, size and ornamentation. Rahman et al. (2019) in their research of living pollen stressed the 'overlapping and diverse morphological characters' in the identification of pollen on the species level.

It seems from the various examples given above and the comparison with living forms in the present study that sculpture alone is not very useful for supra-generic or even generic divisions. Therefore, sculptural elements are not used in the present study as sole criteria for the identification of genera but are considered in combination with some other distinguishing feature, such as the spore outline or nature of tetrad mark. Potonie's (1956) supra-generic classification, namely *infra-turmae* of spores, based on sculptural elements is rejected in the present study for the reasons mentioned above. Dettmann's (1963) supra-generic classification with the exclusion of the *infra-turmae* is adopted for *Sporites*.

Turmae are based on character of the aperture, whether *monolete* or *trilete*. Again, there is a problem with that since one plant may produce spore/pollen with monolete, dilete or trilete aperture, and often there is just a thinning of the exine apparent. Therefore, the term '*Turmae*' is abandoned in the present classification. ***Supra sub-turmae*** are based on the stratification of the *exine*, namely whether the *exine* has resolved into an *ektexine* and an *endexine* by the formation of a cave or cavities between the layers, or whether the exinal layers remain unseparated.

Equatorial features are used in the present classification for sub-*turmae* of the cavate and acavate feature itself; for example, zonate, singulate or singulo-zonate. The character of the zona is used as a criterion for generic diagnosis.

1.5.2 Terminology

Amb; (Ambitus) Erdtman (1952): Outline of spore or pollen grain in polar view.
Apices; The three areas close to the furthest point from the poles in triangular spores.
Brochus–brochi; Erdtman (1952): Mesh of a reticulum, consisting of a lumen and half the muri.
Cavate; Harris (1955): Separation of exine layers to form a cavity or cavities.
Cappa; (cap) Erdtman (1957): In saccate pollen the exine on proximal face of corpus.
Cappula; Erdtman (1957): In saccate pollen, the relatively thin area of exine between sacci roots and distal face.
Cingulum; Potonie and Kremp (1955): An equatorial thickening of the exine in spores.
Commissure; Harris (1955): The line of dehiscence of tetrad scar.
Corpus; Erdtman (1957): The 'body' of a saccate pollen.
Distal face; Erdtman (1952): The face of the spore directed outward in the tetrad, opposite the tetrad mark.
Exine; Erdtman (1952): The main outer, usually resistant layer of the spore or pollen.
Ektexine, sexine; Erdtman (1957): The outer layer of the exine (pollen/spore wall).
Granulate; Potonie and Kremp (1955): Sculptural elements that are more or less round in surface view, and more or less round in side view.
Endexine, nexine; Erdtman (1957): The inner, usually un-sculptured layer of the exine; may be structures.
Laesura-laesurae; Erdtman (1952): The rays of a tri-radiate tetrad scar or the ray of a monolete scar.
Muri; Erdtman (1957): The walls or elevated areas bordering the spaces in a reticulate pattern.
Pilate–pila; Potonie and Kremp (1952): Sculptured elements consisting of a thickened apex and a narrow rod-like neck.
Proximal face; Erdtman (1952): That part of the spore or pollen grain which was directed inward in the tetrad.
Punctate; Erdtman (1952): Minute perforation in the exine.
Reticulate; Erdtman (1952): Sculptural pattern consisting of brochi separated by muri.
Rugulate; Couper (1958): Wrinkled elements irregularly distributed seen as broad rounded crests and troughs in side view and long and narrow in surface view.
Saccus–sacci; Erdtman (1957): The air sac/sacs in saccate pollen grains which is formed by the detachment of the exine layers.
Sculpture; Potonie and Kremp (1955): Elements forming a relief on the surface of the spore or pollen grain and observed in the outline of the spore.

Spines; Erdtman (1952): Long, conspicuous and generally sharp projections. Length exceeding 3 μ.
Striae; Faegri and Iversen (1950): Narrow grooves, more or less parallel (length at least twice the breadth), separated by ridges.
Structure; Potonie and Kremp (1955): Elements that do not form a relief on outer exine surface. Made up of elements present between ektexine and endexine (internal texture).
Sulcus; Erdtman (1952): In pollen grains, a longitudinal aperture or furrow on distal face.
Tetrad; Wodehouse (1935): Union of 4 pollen grains or spores formed by one mother cell.
Taeniae; Jansonius (1962): In saccate pollen grains, the cappa has bands of thickened ektexine divided by grooves.
Tetrad mark; Dettmann (1963): The mark (scar) that is left on the proximal face of the spore, which has been in contact with the other members of the tetrad.
Verrucate; Erdtman (1952): Wart-like projections. Size over 1 μ.
Zona; Potonie and Kremp 1955: An equatorial extension of the ektexine. Differs from the cingulum in not exceeding the thickness of the ektexine which it is part of.

Structural and sculptural terms used when seen in optical section, and surface view follow Erdtman (1952) as well as the illustrations and definitions of Gambarelli et al. (1989).

1.5.3 Bodies Lacking Haptotypic Features: Acritarchs

Alete bodies without any visible haptotypic features are grouped in the present study under *Acritarcha*, a term proposed by Evitt (1963) for 'spore-like bodies of unknown affinity, partly planktonic and probably algal in origin'.

The term 'Acritarchs' refers to single-celled organic microfossils present in rocks from the Cambrian throughout geological history. The earliest in geological time are mostly simple, namely unornamented. The Permian acritarchs range from smooth walled to having ornamentation. They range in size from minute spheres to 100 μ and more. The name 'Acritarcha' originates from Greek, meaning 'uncertain origin'. The name is still used although many acritarchs by now have been assigned to known groups or taxa of aquatic plants. All fossil acritarchs show close resemblance to aplanospores of living algae in the aquatic environment.

Most of the acritarchs described in the present study were scattered in small numbers within various samples throughout the boreholes. However, some acritarchs were restricted to certain stratigraphic levels (see pollen curve charts) and are of great value as indicators of specific environmental conditions at the time. Banerjee and D'Rozario (1990) plotted the distribution of acritarchs in the Lower Permian deposits of eastern India and discussed their significance in interpreting the climatic conditions of the times. Some acritarchs in the present study appear in great numbers in samples otherwise poor in pollen and spores. The spores/pollen in these samples are poorly preserved. Furthermore, these samples were vitrinite-dominated coals poor in exinite macerals, indicating high water table, ideal conditions for wood preservation and the thrive of algae. The few corroded saccate pollen grains found in these samples have come from non-aquatic plants growing on higher ground.

Interestingly, acritarchs are found in fresh water bodies unlike hystrichospheres and later in the dinoflagellates which were dwellers of a marine environment. Li et al. (2004) discussed the different habitats of the two groups. The marine Upper Permian of SE China was dominated by hystrichospheres (Li et al. 2004).

Remains of the alga *Botryococcus* were abundant in only a few samples noted for the absence of acritarchs and abundance of pollen.

Botryococcus is regarded as a fresh water colonial alga. Cookson (1953) mentions that *Botryococcus* usually inhabits fresh water lakes and ponds. However, she also recorded its occurrence in the brackish water lagoons of the Coorong, along the coast of South Australia. I examined coorongite, a rubbery deposit made almost solely of the alga *Botryococcus* deposited in the Coorong lagoon (Glikson 1983; 1984). My study mentions the resistance of *Botryococcus* to degradation and its subsequent excellent preservation in sediments such as the Permian torbanites from the Sydney and Carnarvon Basins. It is noteworthy that *Botryococcus* produces hydrocarbons that are toxic to other algae and plankton, which may explain the absence of acritarchs in *Botryococcus*-rich samples.

1.5.4 Saccate Pollen Grains

In the case of *saccate* pollen grains a vast number of supra-generic groups exist, many of no practical use. Potonie (1955) divided the *saccate* forms into three groups, namely:

1. POLYSACCITES: Cookson (1953)
2. MONOSACCITES: Chitaley (1951)
3. DISACCITES: Cookson (1953)

1.6 Monosaccate Forms

Leschik (1955) further sub-divided the monosaccate forms into two series based on the presence or absence of a proximal tetrad mark. Leschik's series are as follows:

1. TRILETESACCITES
2. ALETESACCITES

The former was proposed by Leschik for monosaccate forms possessing a trilete proximal mark, and the latter for monosaccate forms devoid of a proximal mark.

Bharadwaj (1953) emended vesiculomonoraditi for monosaccate forms and he introduced Araditi (Bharadwaj, 1953) for alete monosaccate forms. In the present study further sub-division of the monosaccites is rejected on the following grounds:

It is evident from many monosaccate forms that the proximal tetrad mark varies from specimen to specimen, from monolete to dilete or trilete, or is absent altogether as a recognisable distinct feature. This morphological feature was traced during the present study by examining over 300 specimens of *Parasaccites gondwanensis* from one sample (see Fig. 51). Therefore, the monosaccate forms in this book are treated as one group, and no further division of the monosaccates is adopted as this would inevitably lead to the splitting of form genera and even species, rendering it useless in comparing microfloras from different sediments of the same age, and thus misleading in stratigraphic correlations.

Hart's (1965) classification of monosaccate forms is based on the *ektexine* mode of detachment, thus sub-dividing the group into two *infra-turmae*:

1. DIPOLSACCITI—are "monosaccites in which the saccus is attached on both the proximal and distal hemisphere of the central body" (Hart 1965).
2. MONPOLSACCITI—are "monosaccites in which the saccus is attached either proximally or distally to the central body, but not on both surfaces" (Hart 1965).

In the present study only one monosaccate form genus seems to be a monpolsacciti; for example, *Densipollenites* Bharadwaj, and therefore Hart's sub-division of the monosaccate forms is not adopted. It is not considered useful to erect a supra-generic group for the sole accommodation of one genus. Nevertheless, Hart's classification in general is adopted on principal grounds; it is based on the consistent feature in a saccate form, namely, the mode of ektexine detachment and formation of the saccus. Interestingly, when fossil saccate pollen grains are compared to living saccate forms the same morphology is valid. On the other hand, the proximal tetrad mark used for sub-division by other authors is an inconsistent characteristic in saccate forms. Furthermore, recent studies of pollen extracted from fossil pollen sacs demonstrated enormous variation in their tetrad mark morphology from mature to immature pollen within the same pollen sac. Furthermore, these monosaccate pollen were classified under Cordaitales, and have been shown to be alete, monolete or trilete. Saccate pollen of living plants show similar diversity in the way of number of sacci (Fig. 1.3a–c).

1.7 Disaccate Forms

The disaccates were further sub-divided by Leschik (1955, 1965) into two series:

1. DISACCITRILETES
2. DISACCI-ATRILETES

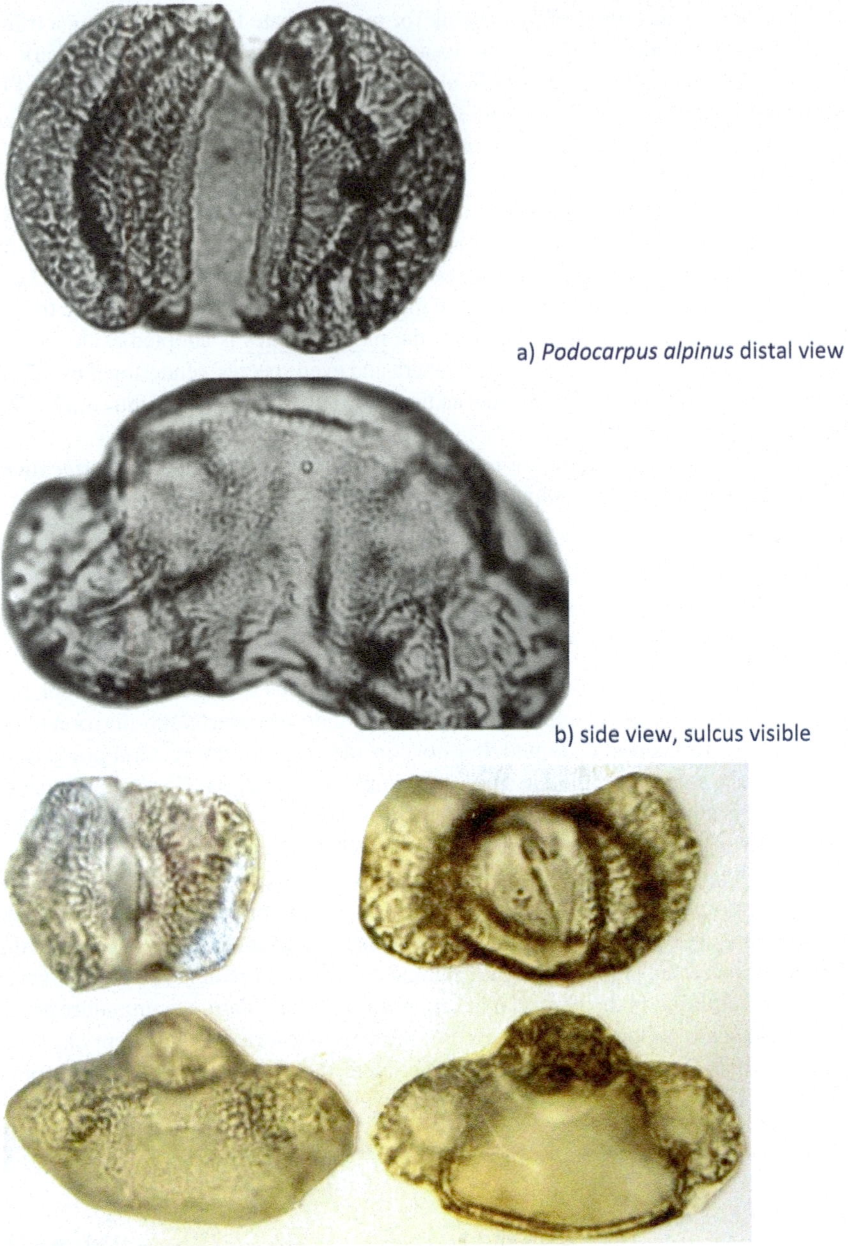

Fig. 1.3 *Podocarpus gracilior*, **a** *Podocarpus alpinus* distal view, **b** side view, sulcus visible

The former was erected to accommodate disaccate forms possessing a trilete proximal mark, and the latter for disaccate forms not possessing one; or possessing a monolete proximal mark. It appears that Leschik (1965, p. 54) interpreted the proximal scar as a germinal aperture. This phenomenon is not known among today's saccate pollen grains which are known to germinate distally. Ancient saccate pollen grains had similar ways of germinating as seen from the studies of cappula which is always a thinner area in the exine, and is also observed in the present study in disaccate pollen.

Again, as in the case of the monosaccate forms, dividing the disaccates on the basis of the characteristics of the tetrad mark would lead to splitting of genera among different supra-generic groups and leading to the formation of artificial species. A fact clearly demonstrated from the studies of the pollen of *Ulmania frumentaria* is described and illustrated by Potonie and Schweitzer (1960). The variations in the tetrad mark in *U. frumentaria* are obvious, passing from monolete to dilete or trilete and from a clearly outlined scar to a vaguely visible one. The same has been observed in the present study of *Potoniesporites* which ranges from monolete to dilete, trilete or inaperturate.

Bharadwaj (1958) erected a new series; SULCATI for sulcate disaccate forms. Series <u>*sulcati*</u> could be used to accommodate such forms as *Falcisporites* Klaus, pollen grains of *Caytonanthus*, which according to Harris (1951) possess "a groove or a slit on the distal face". Series sulcati is not used in the present study simply because only one pollen grain suggesting a sulcate disaccate pollen morphology has been observed in the present samples. It is also noteworthy that no such structure has been found in saccate pollen grains of living plants.

1.8 Striatiti

The proximally taeniate saccate pollen forms are described in the present study under the group STRIATITI (Pant 1955) which is raised here from infra-turma to sub-turma to accommodate all saccate pollen possessing a taeniate cappa regardless of the orientation of taeniae, whether transverse or vertical, and regardless of the character of the sacci, whether disaccate, monosaccate, lobed or rudimentary sacci. The infra-turmae STRIASACCITI and DISTRIASACCITI, proposed by Bharadwaj (1962) for monosaccate forms possessing taeniae on one face of the corpus and on the two faces of the corpus, respectively, are not used in the present study. Monosaccate taeniate forms could possibly be monstrosities of disaccate striatiti, as noted by Hart (1965). The monosaccate or lobed taeniate forms are accommodated here under sub-turma STRIATITI. The striatiti have no equivalent in presently living pollen; they represent an extinct group. Striate pollen has been isolated from sporangia attached to Glossopteris wood. These studies have therefore clearly demonstrated them to be the pollen of the Glossopteris. Moreover, the same studies (Lindstrom et al. 1997) found extreme variations among several thousand pollen grains released from one sporangium; from trisaccate to bisaccate, monosaccate or lacking in sacci entirely.

This fact renders the artificial division of striate pollen into numerous genera and many species based on small variations in size or sacci arrangement obsolete.

References

Backhouse J (1990) Permian palynostratigraphic correlations in south-western Australia and their geological implications. Rev Palaobot Palynol 65:229–237

Backhouse J (1991) Permian Palynostratigraphy of the Collie Basin Western Australia. Rev Palaeobot Pal 67:314–337

Balbach MK (1966) Micrspore variations in Lapidostrobus and comparison with Lycospora. Micropalaeontology 12(3):334–342

Balme BE (1952) The principal microspores of the Permian of Collie. In: Lord JH Collie mineral field. West Aust Geol Surv Bull 105(2):164–201

Balme BE (1959) Some palynological evidence bearing on the development of the Glossopteris flora. Symp Roy Soc Victoria 25:269–280

Balme BE, Hennelly JPF (1955) Bisaccate sporomorphs from Australian Permian coals. Aust J Bot 3(1):89–98

Balme BE, Hennelly JPF (1956a) Monolete, Monocolpate, and Alete Dporomorphs from Australian Permian coals. Aut J Bot 4(1):54–67

Balme BE, Hennelly (1956b) Trilete sporomorphs from Australian Permian sediments. Aust J Bot 4(3):240–260

Banerjee M, D'Rozario A (1990) Palynostratigraphic correlation of Lower Gondwana sediments in the Chuparbhita and Hura Basins, Rajmahal Hills, Eastern India. Rev Palaeobot Palynol 65:239–255

Bharadwaj DC (1953) An approach to the problem of taxonomy and classification in the study of *sporae dispersae*. Palaeobot 4:3–9

Bharadwaj DC (1955) The spore genera from the Upper Carboniferous coals of the Saar and their value in stratigraphical studies. Palaeobot 4:119–149

Bharadwaj DC (1958) On *Porostrobus* zeiller Nthorst and its spores, with remarks on the systematic position of P.bennholdi Bode and the phylogeny of *Densosprites* Berry. Palaeobot 7:67–75

Bharadwaj DC (1962) The miospore genera in the coals of Raniganj Stage (Upper Permian), India. Palaeobot 9:68–106

Brown C (1960) Palynological techniques. Baton Rouge 8, La. 188 pp

Cookson IC (1953) Difference in microspore composition of some samples from a bore at Comaum, South Australia. Aust J Bot 1(3):462–473

Couper RA (1958) British Mesozoic microspores and pollen grains. Palaeontographica 103B:75–179

Dettmann ME (1963) Upper Mesozoic microfloras from southeastern Australia. Roy Soc Victoria, Proc 77(1):148

Eggert DA, Taylor TN (1966) Studies of Palaeozoic Ferns; on the genus Tedelea gen. nov. Palaeontographica 118B:52–73

Erdtman G (1934) An introduction to pollen analysis. Chanica Botanica Comp, Waltham, MA, p 238

Erdtman G (1952) Pollen morphology and plant taxonomy, I Angiosperms, Uppsala, p 539

Evitt WR (1963) A discussion and proposals concerning fossil dinoflagellates, hystrichospheres and acritarchs. Proc Nat Acad Sci 49(3):298–302

Faegri, Iversen (1950) Textbook of modern pollen analysis. Copenhagen

Foster CB (1982) Spore-Pollen assemblages of the Bowen Basin (Australia); their relationship to Permian/Triassic Boundary. Rev Palaeobot Pal 36:1–4

Gambarelli A, Levizzani V, Mandrioli P (1989) Polkey: an expert system for the identification of pollen grains. Aerobiologia 5:17–29
Ghosh SC (2002) The Raniganj Coal Basin: an example of an Indian Gondwana rift. Sed Geol 177:155–176
Glenister BF, Furnish WM (1961) The Permian amonoids of Australia. J Paleont 35:673–736
Glikson M (1983) Microbiological precursors of coorongite and torbanite and the role of microbial degradation in the formation of kerogen. Org Geochem 4(3/4):161–172
Glikson M (1984) Further studies of torbanites and coorongite using transmission electron microscopy and C-isotope analysis. Org Geochem 7(2):151–160
Glikson M (2001) The application of electron microscopy and microanalysis in conjunction with organic petrology to further the understanding of organic-mineral association: Examples from Mt Isa and McArthur Basins, Australia. Int J Coal Geol 47:139–159
Glikson M, Fielding C (1991) The Late Triassic Callide coal measures, Queensland, Australia: coal petrology and depositional environment. Inter J Coal Geol 17:313–332
Glikson M, Golding SD (1998) Organic matter maturation as an indicator of hydrothermsl processes in sedimentary basins. In: Arehart, Hulston (eds) Water-rock interaction. Balkema, Rotterdam, pp 101–104
Glikson M, Owen JAK (1987) A New Ireland coal and associated sediments: hydrocarbon generation from pollen exines at low maturation. J SE Asian Earth Sc 1:221–234
Glikson M, Boreham CJ, Thiede DS (1999) Coal composition, temperature and heating rates; a determining factor in hydrocarbon species generated. In: Mastarelz M, Glikson M, Golding SD (eds) Coalbed methane; scientific, environmental and economic evaluation. Kluwer Academic Press, Dordrecht, Netherlands, pp 257–269
Glikson M, Golding SD, Boreham CJ, Saxby JD (2000) Mineralization in Eastern Australian coals: a function of oil generation and primary migration. In: Glikson M, Mastalerz M (eds) Organic matter and mineralization. Kluwer Academic Publ, pp 314–326
Glikson M, Golding SD, Southgate PN (2006) Thermal evolution of the ore-hosting Isa superbasin: Central and Northern Lawn Hill platform. Econ Geol 101:1211–1229
Glikson M, Hickman AH, Duck LJ, Golding SD, Webb RE (2011) Integration of observational and analytical methodologies to characterize organic matter in Archaean rocks: Distinguishing biological from abiotically synthesized carbonaceous matter structures. In: Golding SD, Glikson M (eds) Earliest life on earth. Springer, pp 209–237
Harris TM (1951) The relationships of Caytoniales. Phytomorphology 1:29–34
Harris WK, Ludbrook NH (1966) Occurrence of Permian sediments in Eucla Basin, South Australia. Geol Surv South Aust Quart Geol Notes 17(11–14)
Hart GF (1965) Microflora from the Ketewaka-Mchuchuma Coalfield, Tanganyika. Geol Surv Tanganyika Bull 36:1–25
Hills ES (1961) Morphotectonics and the geomorphological science with special reference to Australia. Q.J.G.S. 465:77–89
Hughes NF, Dettmann ME, Playford G (1962) Sections of some Carboniferous dispersed spores. Paleontology 5(2):247–252
Jansonius J (1962) Palynology of Permian and Triassic sediments, Peace River Area, Western Canada. Palaeontographica 110B:1–18
Kemp EM, Balme BE, Helby RA, Playford G, Price PL (1977) Carboniferous and Permain Palynostratigraphy in Australia and Antarctica. A review. J BMR Geophys 2:177–208
LeBlanc Smith G (1993) Geology and Permian coal resources of the Collie Basin, Western Australia. West Austr Geol Surv, Rep 38:86
Leschik G (1955) Upper sporen aus dem Salzton des Zechsteins von Neuhof bei Fulda. Palaeontographica 100B:122–142
Li J, Changqun C, Servais T, d'Ascq V, Zhu J (2004) Later Permian acritarchs from Meishan (SE China) in context of Permian palaeontology and Palaeoecology. N Jb Geol Paläont Mh (7):427–448
Lord JH (1952) Collie mineral field. Geol Surv West Austr Bull 105(1):163

Low LG (1958) Collie mineral field. Geol Surv West Austr Bull 105(2):135

McWhae JRH, Playford PE, Lindner AW, Glenister BF, Balme BE (1958) The stratigraphy of Western Australia. J Geol Soc Austr 4:1–141

Mory AJ, Backhouse J (1997) Permian stratigraphy and psalynology of the Carnarvon basin, Western Australia. Geol Surv West Aust 46:1–101

Pant DD (1955) On new disaccate spores from the Bacchus Marsh tillite, Victoria, Australia. Ann Mag Nat Hist Ser 12(8):41–64

Potonie R, Kremp G (1954) Die gettungen der palaontologischen sporae dispersae und ihre stratigraphie. Geol Jb 69:111–194

Potonie R, Kremp G (1955) Die Sporae Dispersae des Ruhr Karbons. Teil I: Palaeoontographica 98B:1–136

Potonie R, Schweitzer NJ (1960) Der Pollen von *Ulmania frumentaria*: Palaeont. Z. 34(1):27–39

Price PL (1983) A Permian palynostratigraphy for Queensland. In: Permian geology of Queensland symposium Brisbane. Proc Geol Soc Austr Queensland Div, pp 155–211

Radforth NSW (1939) Further contribution to our knowledge of the fossil Schizaeaceae; genus senftenbergia. Trans Roy Soc Edinb 59:745–761

Rahman S, Khan SM, Zafar M, Ahmad M, Khan R, Hussain S, Khalid M, Kayani S-I (2019) Pollen morphological variation of Berberis L. from Pakistan and its systematic importance. Microscopy Res Tech 1–8

Remy W (1954/6) Monosaccate Pteridosperm pollen aus dem Karbon und Perm. Palaontl Zeit 28:140–144

Smith AHV (1962) The palaeoecology of Carboniferous peats based on the miospore and petrography of bituminous coals. Proc Yorkshire Geol Soc 33(4):423–474

Staplin FL (1960) Upper Mississippian plant spores from the Golata formation, Alberta, Canada. Palaeontographica 107B:1–40

Taylor GH, Teichmuller M, Davis A, Diessel CFK, Littke R, Robert P (1998) Organic petrology. Bebruder Borntraeger, Berlin, p 704

Teichmuller M (1974) Uber neue Macerale der Liptinit –Gruppe und die Entstehung von Micrinit. Fortschr. Geol. Rheinld. u. Westf. 24:37–64

Wodehouse RP (1935) Pollen grains. In: Their structure, identification and significance in science and medicine. McGrow-Hill, New York, p 74

Chapter 2
Systematic Descriptions

Group: Acritarcha (Evitt 1963).
Genus ***Schizosporis*** (Cookson and Dettmann 1959).
Type species: (original designation) *Schizosporis reticulatus* (Cookson and Dettmann 1959).
Diagnosis: Outline circular to sub-circular. Test splitting equatorially divided into two equal parts. Wall smooth, punctate, granulate or reticulate. No outer membrane is present.
Discussion: *Schizosporis* was first described and defined by Cookson and Dettmann (1959), although similar forms have been recorded and described earlier by a number of authors under various other generic names. The equatorial rupture of the spore test into two equal or almost equal parts places *Schizosporis* in the *Schizomorphitae* which are characterised (Segroves 1967) by an equatorial zone or line of weakness along which it splits readily. Other forms in the group are *Schizocystia* (Cookson and Eisenack 1962), *Circulisporites* (Norris 1965), *Peltacystia* (Balme and Segroves 1966), and possibly *Lecaniella* (Cookson and Eisenack 1962) and *Brazilea* (Tiwari and Navale 1967).

Schizosporis scissus (Balme and Hennelly) (Hart 1960) (Fig. 2.1).
Synonymy: *Laevigatosporites scissus* (Balme and Hennelly 1956).
Schizosporis scissus (Hart 1965a, b) (Fig. 2.1).
Occurrence: Lower to upper Permian of Collie–Muja Basin, Sue-1 bore.
Description: Diameter about 20 μ. Outline circular to sub-circular or oval. Wall one-layered, thin not exceeding 1 μ, smooth on lower test. Upper test has a round thickening which is torn off in greater part in Fig. 2.2b; present in Fig. 2.2a. Test splitting readily equatorially divided into two parts.
Dimensions: Diameter 16–30 μ (20 specimens measured).
Discussion: The sporeis of very small size, and closely resembles a form described by Backhouse (1991) from Collie under the generic name *Brazilea*.

M. Glikson-Simpson, *Coal—A Window to Past Climate and Vegetation*,
https://doi.org/10.1007/978-3-030-44472-3_2

Fig. 2.1 Splitting test

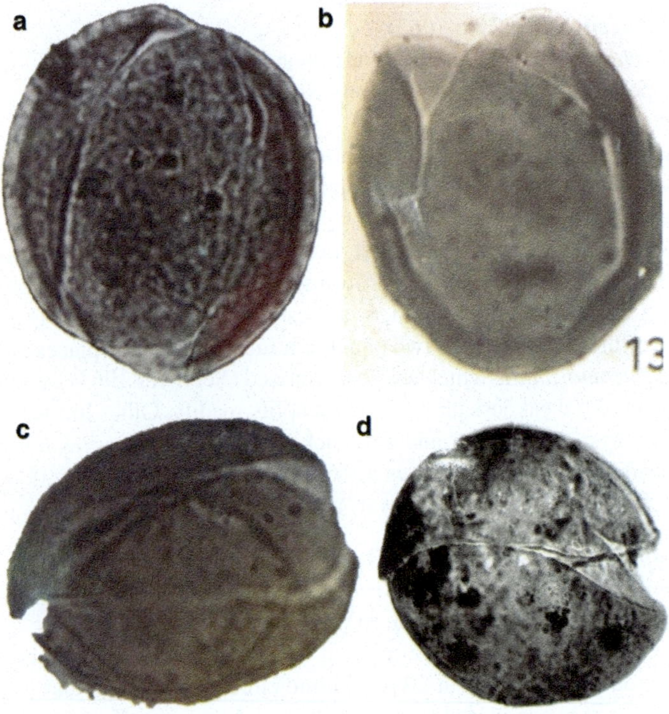

Fig. 2.2 **a**, **b** Partly splitting specimens, **c**, **d** close to splitting in half

Schizosporis balmei (Jansonius 1962) (Fig. 2.2).
Occurrence: Rare, occasionally present in Upper Permian samples of Collie and Wilga Basins.
Description: Outline circular to sub-circular, smooth. Wall one-layered 2–3 μ thickness and smooth. Test splitting equatorially. It is significantly larger than *Schizosporis scissus*. That fact alone does not justify treating the spore as a separate species. However, it does not appear in the same stratigraphic horizon as *S. scissus*.
Dimensions: 65–105 μ in diameter (10 specimens measured).
Remarks and affinity: *Schizosporis* as the name implies is a spore that splits upon germination. Spores closely resembling *Schizosporus* today are those of *Equisetum fluviatile* Fig. 2.3a–e, which implies a higher water level, above the muddy swamp. These spores are excellent indicators of environmental conditions at the time of deposition. A similar acritarch has been described by Li et al. (2004) from the Late Permian of South East China.

Genus ***Peltacystia*** (Balme and Segroves 1966).
Description: Outline circular to sub-circular. Wall one-layered, sculptured with concentric ridges and processes. Test splitting equatorially along a zone of weakness.
Known distribution: Upper Permian of northern Perth Basin, Western Australia, Lower Permian of Eastern India (Banerjee and D'Rozario 1990), Permian of Collie Basin (Backhouse 1991).
Discussion: *Peltacystia* is distinguished from other members of the **Schizomorphitae** by its unique sculpture of circum-polar ridges which in turn are covered in flat-topped spines.
Peltacystia like *Schizosporis* splits equatorially (Fig. 2.4).

Peltacystia venosa (Balme and Segroves 1966) (Fig. 2.4).
Occurrence: Lower Permian to Upper Permian of Collie–Muja Basin, more frequent in Upper Permian samples than in the Lower Permian.
Description: Entire specimens spheroidal to sub-spheroidal. Usually only circular to sub-circular halves are found. Wall one-layered and sub-divided into thicker central area and a thinner marginal zone. Sculpture of outer surface of wall is complex and consists of concentric thickenings or ridges which support the protrusions of the major sculptural elements. These protrusions extend to about 2 μ away from the surface. In between the concentric ridges which numbered 1–3, low-relief ridges or veins are present. The lower ridges are arranged in a polygonal or sub-reticulate pattern. The ruglate-pilate processes project from the corners of the polygons at fixed intervals. The processes which widen at their bases and slightly at their tips are a composite of a regulate and pilate sculpture.
Dimensions: 38–70 μ in diameter (15 specimens measured).
Remarks: Algal spores resembling *Peltacystia* such as species of *Pediastrum* are found today in fresh water lakes.

Peltacystia monile (Balme and Segroves 1966) (Fig. 2.5).
Occurrence: Artinskian to upper Permian of Collie–Muja.

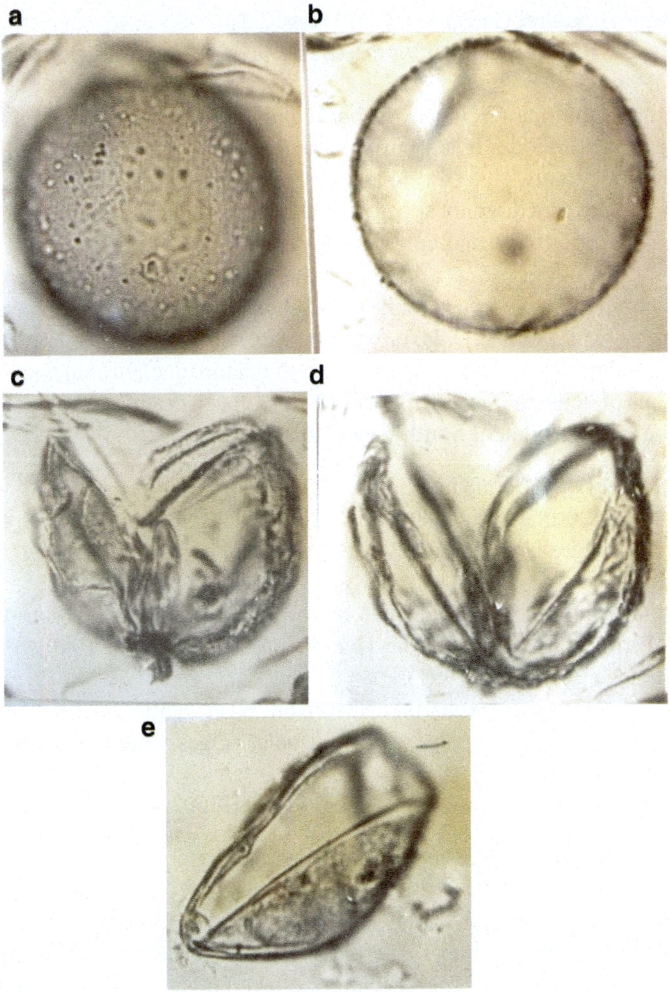

Fig. 2.3 A *Equisetum fluviatile*; immature pollen, **a** whole pollen. **b**, **c** Mature pollen in the process of splitting, **d** split pollen of *Equisetum*

Description: Whole specimen spheroidal. Test splitting equatorially into two equal parts. The test of each hemisphere bears a circumpolar ridge of thickened wall.
Dimensions: 25–35 μ (8 specimens measures).
Affinity: The fern spore *Hypolepis repens* growing in Australia today is the closest in comparison to *Peltacystia monile*. It also splits equatorially and has intricate sculptured exine (Fig. 2.5).

Genus ***Circulisporites*** (de Jersey 1962; Norris 1965).
Diagnosis: Test spheroidal. Equatorially splitting into two equal parts. Wall monolayered. Sculpture consists of concentric ribs.

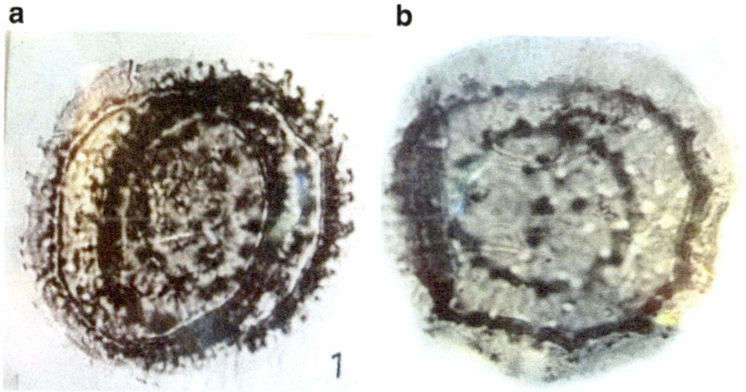

Fig. 2.4 **a** A test split into two halves. **b** One half of a test

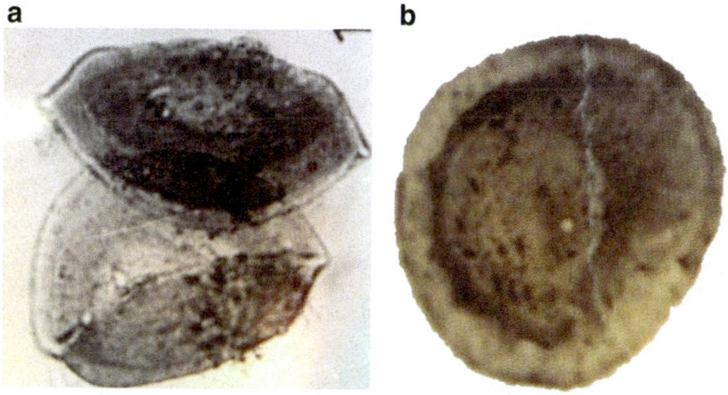

Fig. 2.5 **a** Test splitting into two parts, **b** one half of a test

Distribution: Widely distributed and reported from many parts of the world. In Australia was described and reported from the Permian of Collie Basin (Balme 1952), from the Devonian of Carnarvon (Balme 1959), from the Triassic of Tasmania (Playford 1965), Triassic of Victoria Land, Antarctica (Norris 1965), Permian of India (Banerjee and D'Rozario 1990), Tertiary of New South Wales (J. A. Owen, pers. comm.).

Circulisporites parvus (de Jersey 1962; Norris 1965) (Fig. 2.6).
Occurrence: Sakmarian to Upper Permian of Collie–Muja Basins. Frequent in the following samples: Collie Site B bore 664′-68′, 800′.
Description: Outline circular. Whole tests spheroidal, very small 23 μ in diameter. Wall one-layered, less than 1 μ in thickness. Sculpture ribbed; ribs closely spaced, faintly visible, of low relief, arranged in a concentric pattern.
Dimensions: 18–20 μ (20 specimens measured) (Fig. 2.6).

Sub-group Sphaeromorphitae (Downie, Evitt and Sargeant 1963)
Genus ***Verrucosphaera*** gen. nov.

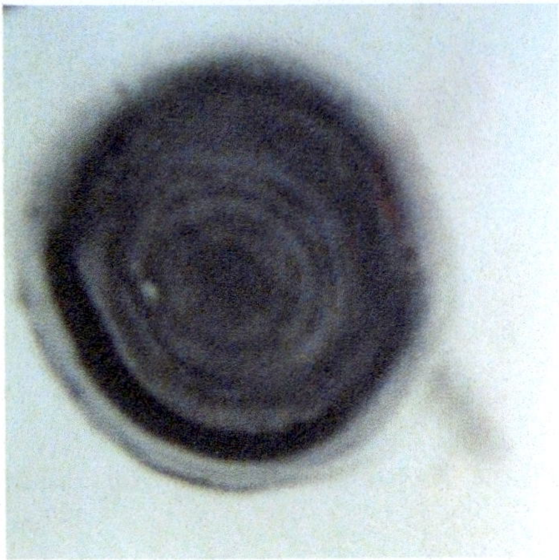

Fig. 2.6 General view of *Circulisporites*

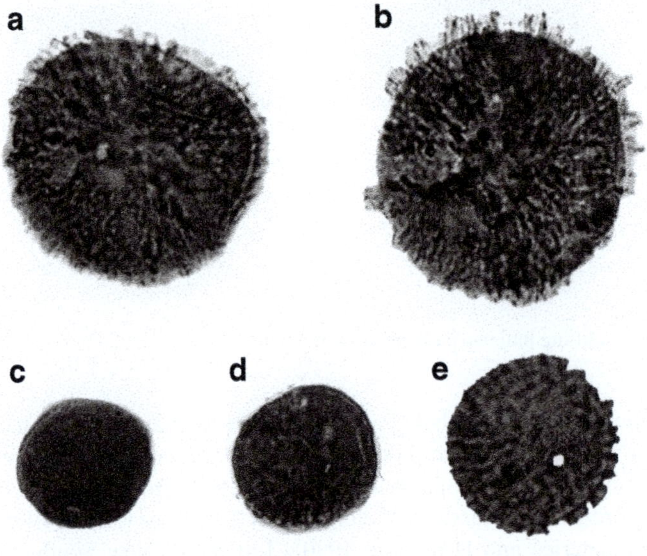

Fig. 2.7 **a**, **b** Mature spores, **c** immature spore, **d**, **e** immature spore with beginnings of sculpture

Verrucosphaera colliensis sp. nov (Fig. 2.7).
Diagnostic features: Outline circular. Wall two-layered; inner wall up to 2 μ thick, overlain by a sculptured outer membrane which is devoid of sculpture in immature spores. Sculpture starts developing with spore maturation (Fig. 2.7c). Sculpture appears as verrucate, pilate or micro-bacculate in surface view; individual elements about 1 μ long or longer in mature spores. A pylome is present in immature, semi-mature and mature spores.
Specimen: Holotype U.W.A. 61925. Isotype U.W.A. 60941.

Type locality: Western Australia, Collie Basin, Site B Bore, 800′.
Age and occurrence: Upper Permian of Collie and Muja Basins. Abundant in vitrinite-dominated coals.
Dimensions: 24–60 μ (30 specimens measured).
Discussion: A significant number of *Verrucosphaera* specimens show a small, circular pylome-like opening. Sculpture small, underdeveloped and not covering entire spore in immature specimens. On the other hand, sculpture becomes denser in mature spores and larger specimens where projections may reach 3 μ. Algal spores closely resembling *Verrucosphaera* are found today in fresh water ponds.

Genus ***Cymatiosphaera*** (Wetzel 1933; Deflandre 1954)
Diagnosis: Spherical or ellipsoidal. Wall two-layered. Outer membrane divided into polygonal fields. Inner wall smooth. Corners of polygons are generally thickened and elevated above spore surface.
Discussion: *Cymatiosphaera* was first described from the Cretaceous of Germany (Wetzel 1933). The main characteristic of the form as described by Wetzel was the possession of a "lamellar skin, which is supported by rods". Deflandre (1954) interpreted the outer layer as a membrane arranged in a polygonal pattern with thickened angles which appear as rods when observed in lateral view. The reticulate or polygonal membrane distinguishes *Cymatiosphaera* from other acritarchs. Wetzel (in Norris and Sarjeant 1965) suggested a similarity to the radiolarian *Dictyospyris* based mainly on the reticulate pattern of the outer membrane. However, *Cymatiosphaera* is made of organic matter and therefore not a radiolarian. Tiwari (1964) described the same structured acritarch designating it to *Maculatasporites*; having sub-spherical test wall, one-layered, made up of mesh work, suggesting a reticulate pattern, consisting of muri only. The same applies to *Cymatiosphaera*; characterised by its mono-layered wall made up of an open mesh of a reticulum. A thin membrane that is only preserved in some specimen and missing in many others is a distinct feature. Both morpho-types are present in the same samples. The wall structure of *Maculatasporites* resembles members of the Linotolypydae family (Eisenack 1962). However, the difference is in the structure of the muri which in Linotolypydae is made of fibres unlike the solid muri in *Maculatasporites/Cymatiosphaera*.

Cymatiosphaera gondwanensis (Tiwari) New Comb (Fig. 2.8).
Synonym: *Maculatasporites gondwanensis* (Tiwari 1964).
Occurrence: Lower to Upper Permian of Collie; Site B bore 1641′; Muja Site D bore 665-675′.
Description: Spherical to sub-spherical. Reticulate inner layer when membrane present. Membrane intra-punctate to intra-granulate. In specimen where membrane is missing a hollow reticulum makes up the wall of the spore. Muri about 2 μ thick surrounding a void 6–8 μ across.
Dimensions: 45–80 μ across (20 specimens measured).

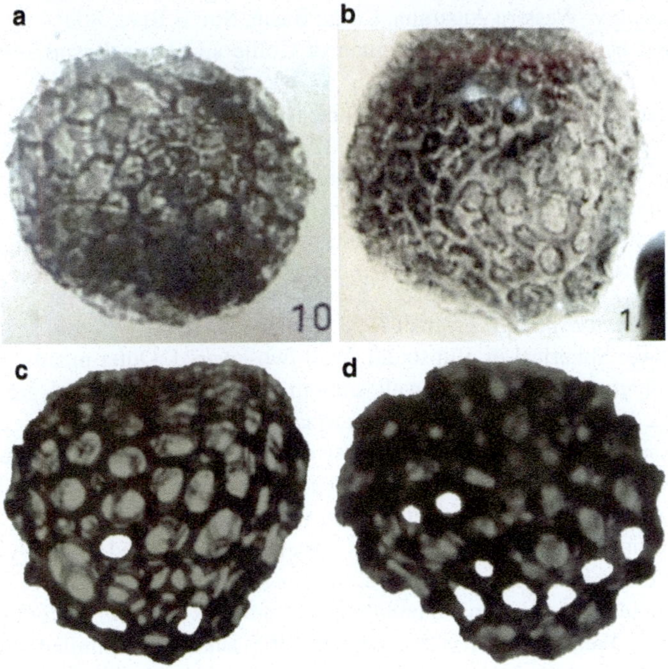

Fig. 2.8 **a** Spore having ektexinal layer, **b** spore with ektexinal layer, **c**, **d** 'naked' spores, ektexinal layer absent. Spores have germinated at some stage

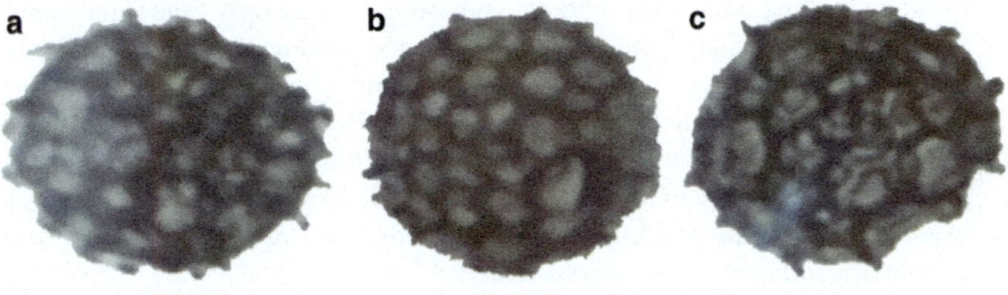

Fig. 2.9 **a–c** *C. microreticulata*

Discussion: As mentioned in the generic description, *Maculatasporites* and *Cymatiosphaera* are one and the same genus, being aplanospores of the same alga. Open reticulum of exine signifies germination of the spore at some stage (Fig. 2.8).

Cymatiosphaera microreticulata sp. nov (Fig. 2.9).
Occurrence: Lower Permian and Upper Permian of Collie Basin. Type Locality: Site B bore, 1641′.
Specimen: Holotype; U.W.A. no. 61888. Isotype; U.W.A. no. 61887.

Description: Small, about 20 μ in diameter. Spherical test with a smooth inner wall 1–1.5 μ thick. Outer membrane thin and structured—reticulate. Reticulum consists of muri 1.5 μ in thickness surrounded by brochi of up to 4 μ across. A small rounded germination aperture is apparent in some specimen.
Dimensions: 15–30 μ (mean 21 μ) (30 specimens measured). Known distribution: Lower Permian of India (Tiwari 1964).
Discussion: The minute size and smooth membrane of this acritarch are its distinguishing feature. The membrane enveloping the spore in *C. gondwanensis* is intra-punctate to intra-granulate which is the structure that distinguishes it from *C. microreticulata*.

Sub-group **Acanthomorphitae** (Downie, Evitt and Sarjeant 1963).
Genus ***Mehlisphaeridium*** (Segroves 1967).
Test spheroidal. Two-layered wall; sculptured as well as structured. Sculpture consists of hollow and differentially thickened processes. Structure infra-punctate.
Known distribution: Artinskian to Upper Permian of the Perth Basin (Segroves 1967), Permian of Collie Basin (Backhouse 1991).

Mehlisphaeridium fibratum (Segroves 1967) (Fig. 2.10).
Occurrence: Lower Permian (Artinskian?) to Upper Permian of Collie Basin, Site B bore 1641′, 224-27′.
Description: Spheroidal test. Two-layered wall. Inner wall solid, rather thick, up to 2 μ, overlain by an outer transparent membrane. Membrane sculptured and structured. Structure is infra-punctate. Sculpture consists of numerous pseudo-echinate processes, varying in length from 4 to 20 μ and tapering at their ends. Membranous processes are reinforced by columnar thickenings of the membrane (see reconstructed specimen).
Faint indication of vein structure can be sometimes observed in the linear arrangement of the structural elements along the thickenings. The thickenings vary in number in one process from three to five. The total number of processes varies from 10 to more than 20 in different specimen.
Dimensions: 30–50 μ in diameter (20 specimens measured) (Fig. 2.10).

Genus ***Tetraporina*** (Naumova 1953, 1961).
Diagnosis: Test quadrilateral. Apices rounded. Wall one-layered, smooth to granulate. Arcuate folds may or may not be present around apices.
Known distribution: Lower Carboniferous of Moscow Basin (Naumova 1953, 1961), Lower Carboniferous of Spitzbergen (Playford 1965), Carboniferous of Canada (Staplin 1960), Lower Permian of Victoria (Pant and Mehra 1963), Lower Triassic of Western Australia (Balme 1963), Carboniferous of Saudi Arabia (Hemer and Nygreen 1967).
Discussion: Playford (1965) discussed the affinity of *Tetraporina* and suggested an algal origin. Scott, Barghoorn and Leopold (1960) also noted its resemblance to the unicellular green alga *Tetraedron*. Churchill (1960) recovered forms closely resembling *Tetraporina* from recent and tertiary deposits in Western Australia.

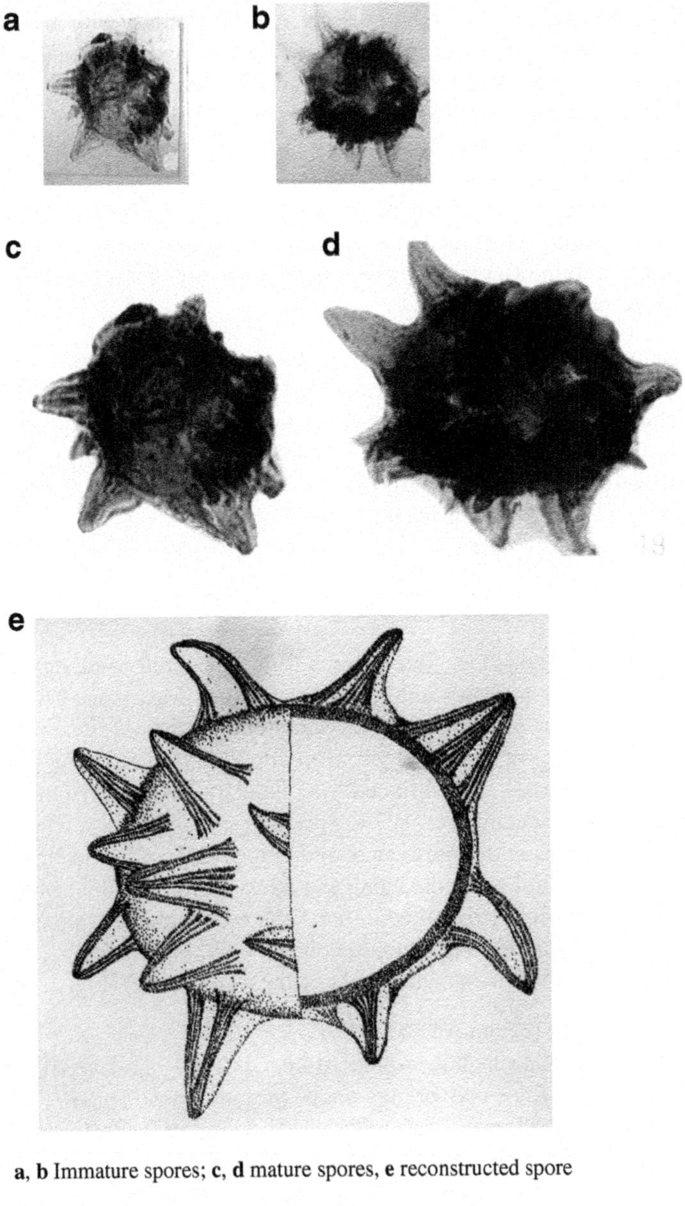

Fig. 2.10 **a**, **b** Immature spores; **c**, **d** mature spores, **e** reconstructed spore

Tetraporina horologia (Staplin) (Playford 1965) (Fig. 2.11).
Occurrence: Sakmarian to Upper Permian of Collie–Muja Basin, Wilga Basin, frequent in the following samples: Collie Site B bore 2513′, 1641′; Muja Site D bore 201-212′. Upper Permian of Alexandra Bridge bore; core 2; 1505-1515′.
Description: Outline quadrangular. Sides straight to concave; usually two sides are straight while the opposite two sides may be more or less concave. Most

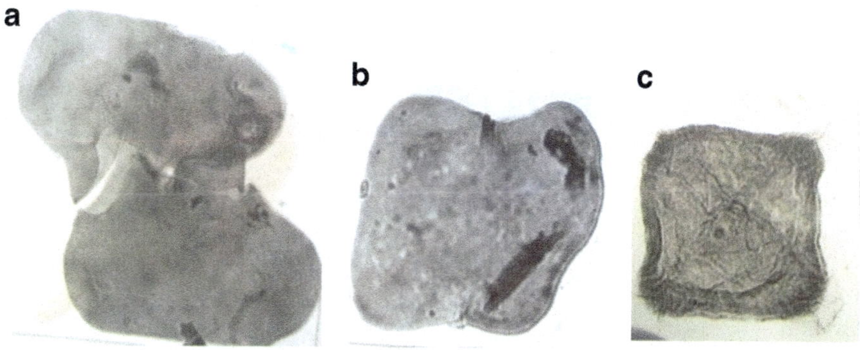

Fig. 2.11 **a** Mature spore, split. **b** Immature spore. **c** Immature spore with membranous overlay

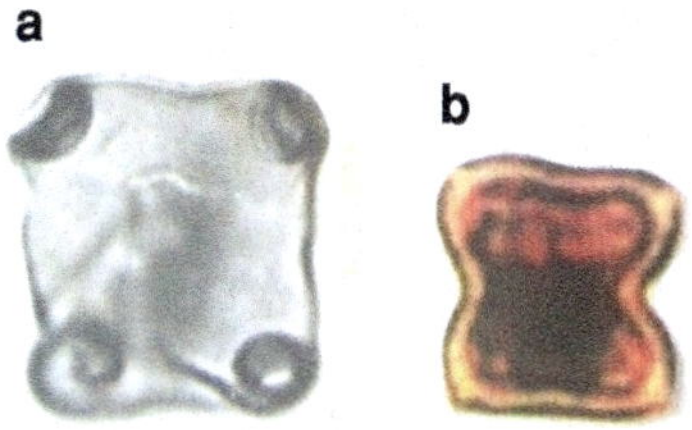

Fig. 2.12 **a** Permian *Tetraedron*. **b** Living alga *Tetraedron*

specimens encountered are mono-layered; exine smooth. Occasional specimen was encountered with a membranous ektexine enveloping the smooth thicker endexine. Membrane is fitting along corpus and loose over apices. The significant size range is attributed to different maturation stages of the spores belonging to the same alga.
Dimensions: 70–135 μ (35 specimens measured).

Cf. ***Tetraedron minimum*** (Hansgirg 1888).
Occurrence: Sakmarian to Upper Permian of Collie–Muja Basin and Wilga Basin.
Description: Small, quadrangular shape. Distinct thickenings of exine around apices.
Remarks and comparison: cf. *Tetraedron* appears identical in morphology to the living alga of the same name. *Tetraedron* (Fig. 2.11c) was collected from the south-eastern part of the Baltic Sea and photographed by Olenina in 2006 for the Scandinavian Culture Collection of algae, the University of Copenhagen (Figs. 2.12 and 2.13).

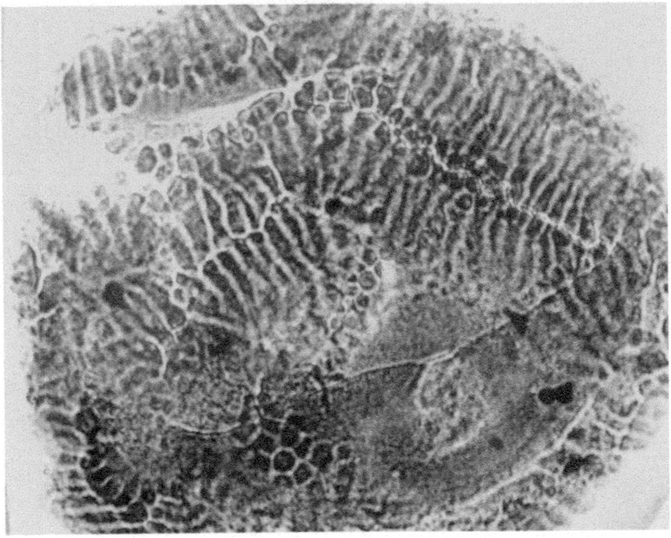

Fig. 2.13 *Botryococcus* algal remains

Botryococcus braunii (Kützing) (Blackburn 1936) (Fig. 2.13).
Occurrence: Very rare. Muja No. 1 Bore; 1126-1130′ Remarks: *Botryococcus* is found in fresh to brackish water. Highly resistant to degradation (see discussion in Part I).
Anteturma: **Sporites** (Potonie 1893).
Turma: **Triletes** (Potonie and Kremp 1954).
Subturma: **Perinotriletes** (Dettmann 1963).
Genus ***Zinjisporites*** (Hart 1963).
Diagnosis: Cavate. Heavily ornamented trilete spores. Amb. triangular to rounded-triangular. Laesurae usually indistinct. Sculpture consists of baculae which coalesce at their bases to simulate a zona.
Known distribution: Lower Permian (Artinskian) of Tanganyika (Hart 1963). Similar forms were described by Balme and Hennelly from the Permian of New South Wales under *Verrucosisporites bullatus*.
Remarks: The description of *Zinjisporites* by Hart conforms with spores found in Collie–Muja. This spore is so abundant in some samples so as to yield good specimens for a detailed description. The cavate appearance is often obscured by the heavy sculpture of the ektexine.

Zinjisporites eccensis (Hart 1963) (Fig. 2.14).
Occurrence: Lower Permian (Artinskian?) of Collie–Muja Basins; Site B bore/1641′; Muja Site D bore/1847′.

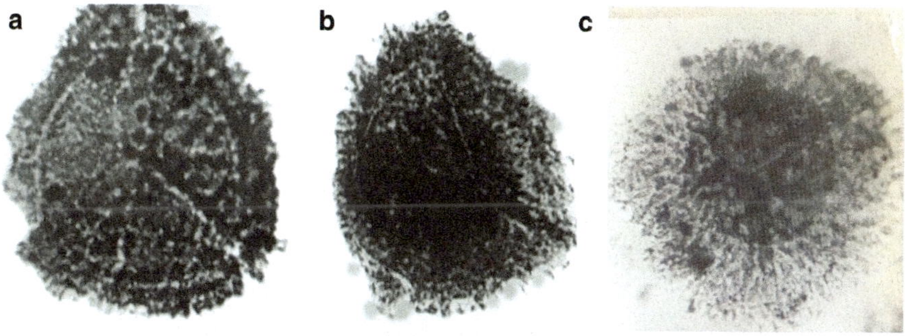

Fig. 2.14 a–c Proximal view showing highly dense sculpture obscuring laesurae and spore outline. Cavate form just visible

Description: Cavate. Trilete. Amb triangular, rounded-triangular to sub-circular. Laesurae not distinct but when visible long and extending to equatorial margin of the spore. Sculpture of ektexine complex; essentially composed of long baculae to pilate processes which fuse frequently at their base. Sculptural elements densely spaced, irregular in size; their length varies according to the state of preservation, often with broken ends.
Dimensions: 33–60 μ in equatorial diameter (30 specimens measured).
Reconstructed spore showing the distinct cavate morphology of the spore and the interconnected projections making the sculpture.
Figure showing irregularity of sculptural elements.

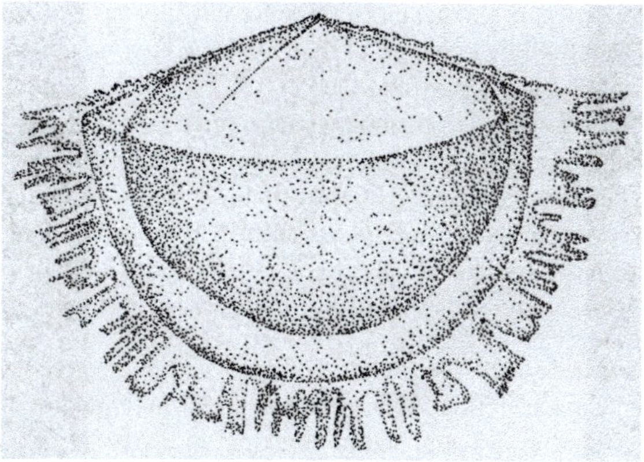

Zinjisporites cf. *bullatus* (Fig. 2.15)
Occurrence: Rare, only less than a dozen specimens were observed, coming from different samples, ranging in age from Sakmarian to Upper Permian of Collie, Site B bore.

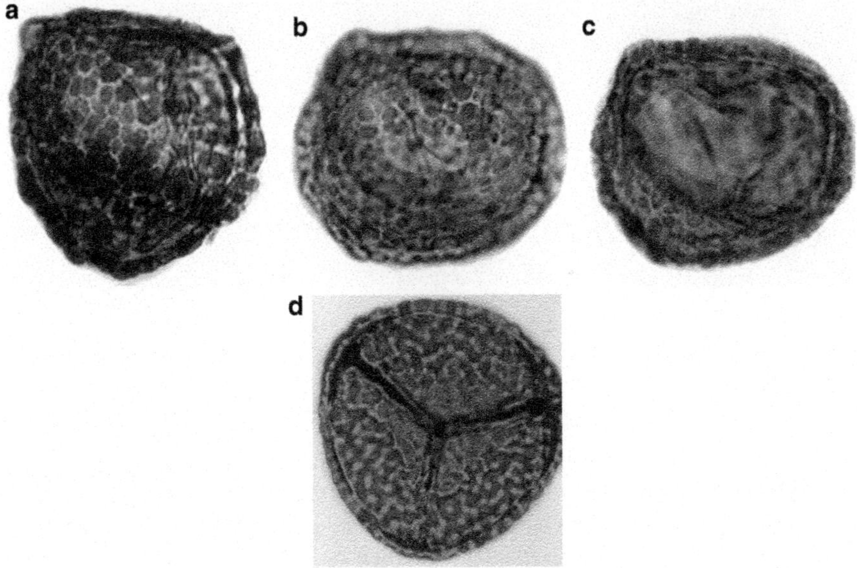

Fig. 2.15 **a** Distal view showing verrucate sculpture. **b** and **c** proximal face mostly lacking sculpture. **d** *Lycopodium sp*

Description: Cavate. Trilete. Amb sub-triangular to sub-circular. Laesurae indistinct. Endexine hyaline, smooth, attached on proximal face and detached from ektexine on distal face.

Ektexine verrucate: Verrucae of low relief, closely packed and varying in shape and size. Mostly they coalesce at their bases. Ektexine up to 3 μ thick on distal face and about 1 μ on proximal face.

Dimensions: 45–48 μ (6 specimens measured).

Remarks and comparison: *Zinjisporites* sp. Differs from *Z. eccensis* and *Z. zonalis*, both described by Hart (1963) from the Permian of Tanganyika, by its verrucate sculpture. *Zinjisporites bullatus* described by Balme and Hennelly from the Permian of New South Wales is characterised by its bulbous processes.

Resemblance to the fern spore of *Lycopodium sp.* is (Fig. 2.15d) noticeable.

Genus ***Densoisporites*** (Dettmann 1963).

Diagnosis: Cavate. Trilete. Ektexine structured finely and thickened equatorially. Endexine (inner layer) attached to ektexine on proximal face only; thin, smooth except for inter-radial papillae.

Discussions: *Densoisporites* was originally described by Weyland and Krieger (1953) from the Cretaceous of Germany, as cingulate, trilete spore, without mentioning its cavate character. Dettmann (1963) emended the genus, emphasising its cavate exine, the cingulum and the fine structure of the ektexine, as well as the apical papillae of its endexine.

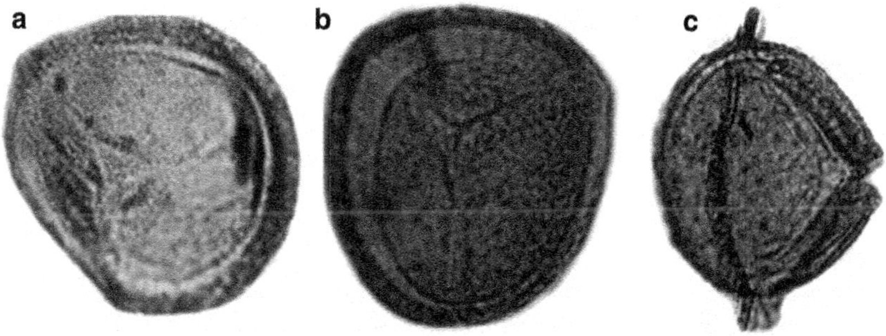

Fig. 2.16 **a** and **b** Proximal view showing pronounced cavate characteristic. **c** Side view with cingulum

Densoisporites sp (Fig. 2.16).
Occurrence: Upper Permian of Collie Basin.
Description: Trilete, cavate and cingulate. Cingulum hyaline, almost transparent, occasionally torn off. Amb. rounded to triangular rounded. Laesurae simple, long, extending to or almost to equatorial margin of spore. Exine cavate; endexine smooth, thin attached on proximal face and detached on distal face. Ektexine about 1 μ thick, thinning out towards proximal face, and thickening equatorially to form a narrow cingulum about 2 μ wide. Structure of endexine dense, intra-granulate; grana closely packed. Structure fades out towards proximal face, becoming faint over contact area.
Dimensions: 40–50 μ (12 specimens measured).
Remarks: *Densoisporites* described by Backhouse (1991) from Collie Basin has a clearly defined thicker cingulum. *D. sp.* has a very fine cingulum, often only a rudimentary one. On the other hand, *D. sp* is characterised by a pronounced cavate appearance. A cavate presence is not visible in the *Densoisporites* described by Backhouse. Furthermore, the laesurae in the present specimens of *Densoisporites* are not accompanied by folds; when present point to elevated laesurae before compaction.

Genus ***Indotriradites*** (Tiwari) (Foster 1979).
Synonym: *Kraeuselisporites* (Leschik 1955).
Diagnosis: Zonate, cavate spores. Trilete; laesurae often extending onto flange in the form of thickenings. Commissures simple, raised. Endexine thin and smooth. Ektexine structured and sculptured. Structure intra-punctate. Sculpture granulate, echinate and usually restricted to the distal face. If extending onto proximal face, elements noticeably reduced in size. Flange smooth or sculptured.
Known distribution: Lower Triassic of Canada (Jansonius 1962), Lower Triassic of Western Australia (Balme 1963), Permian of New South Wales (Balme and Hennelly 1956), Triassic of Tasmania (Playford 1965), Lower Coal Measures of Tanganyika (Hart 1960), Lower Permian of the Barakar coal in India (Tiwari 1964, 1965).

Discussion: *Indotriradites* (Tiwari 1964) closely resembles *Kraeuselisporites* which is treated as its synonym.

Indotriradites is used in the present study to accommodate cavate and zonate spores with a characteristic-coarse echinate/spinate sculpture on the distal face which extends to the cingulum and gives it a spinate outline in optical section. Spines can be very dense giving the spore a thick dark look.

Indotriradites splendens (Balme and Hennelly) (Foster 1979).
Synonymy: *Cirratriradites splendens* (Balme and Hennelly 1956).
Occurrence: Lower to Upper Permian of Collie–Muja Basin; bore B/973-93′.
Description: Cavate and zonate. Amb triangular. Trilete; scar distinct, laesurae extending usually onto flange in the form of exinal thickenings. Commissures usually raised. Structure of ektexine intra-punctate. Sculpture echinate, confined to distal face, consists of 4–10 μ long spines, with sharp ends, widening towards their bases to 4–10 μ in diameter. Spines are closely spaced, about 1 spine on a 4 sq. μ area; they extend onto flange, although usually in a reduced size. Flange uniform in width all around spore equator except for a slight widening around apices, varying in width from 10 to 20 μ. A dark ring of variable width present at flange base, representing the distal face infold and the flange base.
Dimensions: 48–60 μ (20 specimens measured).
Remarks and comparison: *I. splendens* is distinguished by its closely spaced, wide-based coarse spines confined to its distal face and its intra-punctate structure. *Indotriradites korbaensis* (Tiwari 1964) has slightly less coarse sculpture, otherwise closely resembles *I. splendens* and may well be its synonym. The spores are distinct from *Indotriradites niger* by the coarse, densely spaced, wide-based echinate sculpture that often forms a pseudo-cingulum in *I. splendens.*

Indotriradites niger (Backhouse 1991).
Diagnosis: Cavate and zonate. Trilete. Amb triangular to rounded-triangular. Laesurae long. Structure intra-punctate. Sculpture echinate, fading towards the proximal face and towards flange.
Occurrence/ type locality: Lower to upper Permian of Collie–Muja Basins. More common in the Lower Permian. Collie Basin site B bore-1980′.
Description: Cavate and zonate. Trilete; laesurae long, extending to equatorial margin of spore and often onto flange. Amb triangular to rounded-triangular. Exine cavate and zonate; zona 8–10 μ wide. Structure of ektexine dense, fine intra-punctate. Sculpture echinate. Spines long and often broken at their tips giving the sculpture a verrucate appearance (Fig. 2.17a).
Dimensions: 45–66 μ (30 specimens measured).
Remarks and comparisons: *I. niger* is distinguished from most of the other similar forms of the same genus by its fine, intra-punctate structure, and sparsely distributed echinate sculpture which contrasts with the coarse dense spaced echinate sculpture of *I. splendens.*

I. niger was described and designated by Backhouse (1991) from the Collie Basin. The same author points out the difficulty of assigning the species due

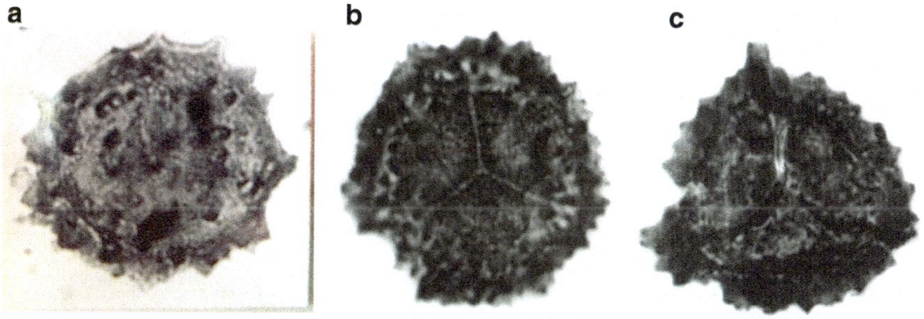

Fig. 2.17 **a** Distal view. **b** Proximal view. **c** Echinate cingulum, a characteristic of this species

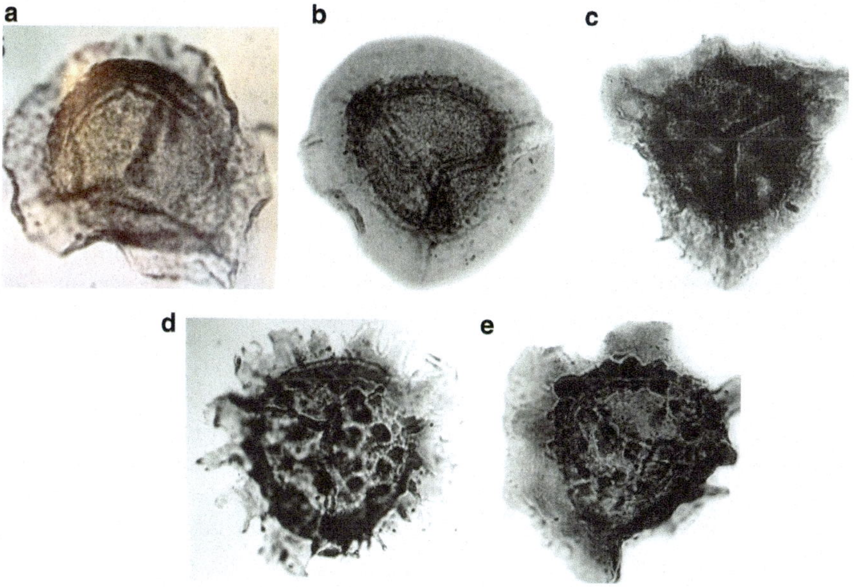

Fig. 2.18 **a** General view showing cingulum and folds along laesurae and intra-punctate structure (immature spore). **b** Structure and reduced sculpture in immature spore. **c** Mature spore with well-developed echinate sculpture and wide cingulum. **d** Mature spore; wide bases of spines and wide cingulum, torn in parts. **e** Mature spore; spine tips broken, only bases left which have the appearance of pseudo-verrucate sculpture. Immature spore

to the immense variability of the sculpture, as also evident from the present study (Fig. 2.17). Immature spores display subdued sculpture while mature spores are characterised by coarse sculpture and a wide cingulum, which is often torn (Fig. 2.18).

Genus ***Gondisporites*** (Bharadwaj 1962).
Diagnosis: Cavate and zonate. Trilete; laesurae long. Amb rounded-triangular. Ektexine sculptured and structured. Sculpture echinate. Structure intra-punctate

to infra-granulate. Zona of uneven width, relatively narrow, often restricted to merely a ridge.
Known distribution: Upper Permian of India (Bharadwaj 1962).
Discussion: *Gondisporites* was named by Bharadwaj for trilete spores with a poorly developed zona. Bharadwaj noted an 'inner body' which appears to be the nexine of a cavate spore. The uneven and usually narrow flange in *Gondisporites* and sparsely dispersed spines distinguish it from *Indotriradites*.

Gondisporites raniganjensis (Bharadwaj 1962) (Fig. 2.19).
Occurrence: Upper Permian of Collie–Muja Basins.
Description: Cavate, zonate. Trilete; laesurae simple, long, usually extending to inner margin of flange. Amb circular to rounded-triangular. Proximal face pyramidal. Distal face hemi-spherical. Exine thinner on proximal face than on distal face. Sculpture echinate, confined to distal face; spines are 1.5–3 μ long, and 3.5–4.5 μ long. Flange irregular in width in a single specimen.
Dimensions: 60–75 μ in equatorial diameter (15 specimens measured).
Remarks and comparisons: *G. raniganjensis* is distinguished by its coarse spinnate sculpture; spines, though a large and a dominant feature are spaced apart. Flange is wider than in other species of *Gondisporites*. *G. raniganjensis* is significantly larger than *Gondisporites sp*.
Gondisporites raniganjensis is an important indicator spore that marks the Late Permian in Collie Basin as well as other Gondwana Basins, such as the Late Permian coals of India (Tripath et al. 2012) (Fig. 2.19).

Indotriradites surangei (Tiwari 1965) (***Gondisporites sp***.) (Fig. 2.20).
Occurrence: Upper Artinskian of Collie–Muja Basin (Collie Basin Bore B/1468′).
Description: Trilete, cavate. Outline circular to rounded-triangular. Distal face hemi-spherical. Laesurae of variable length. Zona narrow, usually forming only an uneven, equatorial ridge. Structure of ektexine intra-punctate. Sculpture echinate, confined to distal face. Length of spines usually no more than 1 μ in length, and 2–3 μ apart.
Dimensions: 40–85 μ in equatorial diameter (30 specimens measured).
Remarks: *Indotriradites surangei* was described by Banerjee and D'Rozario (1990) from the Permian of Eastern India. The same authors used this species as one of the most important vegetational climatic indicator, in combining palynological zones with climatic and ecological phases. Their placing fits well with that of Collie Basin. *I. surangei* enters the sedimentary sequence in the Collie Basin in middle Artinskian, which corresponds to Banerjee and D'Rozario's PNH5 zone. It is therefore one of the environmental indicators of fern spores. In Appendices D and E it is under *Gondisporites* sp. There is a close morphological resemblance between the two genera. However, *Gondisporites raniganjensis* appears earlier than *Indotriradites surangei* in the stratigraphic column (Appendices D and E).

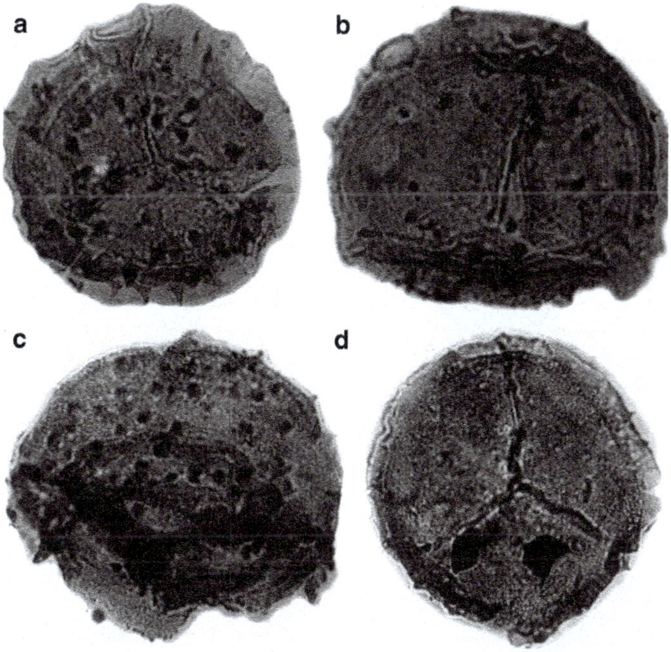

Fig. 2.19 **a** Proximal face, showing echinate sculpture and intra-punctate structure. Cingulum smooth, devoid of structure. **b** Semi-distal face. **c** Side view, showing absence of cingulum on distal face. **d** Proximal face showing intra-granulate structure

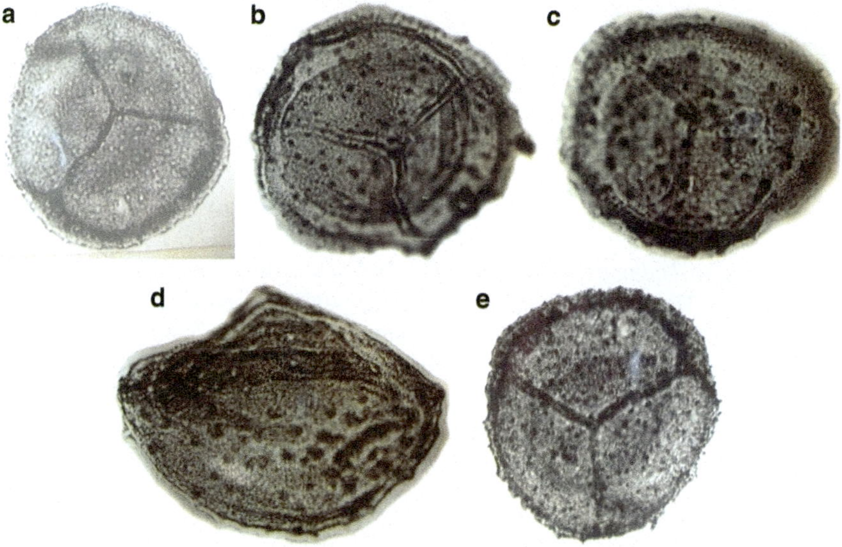

Fig. 2.20 **a**, **b** Proximal face, **a–e** cavate character of spore clearly visible; long trilete scar prominent. **e** Proximal and partial distal view

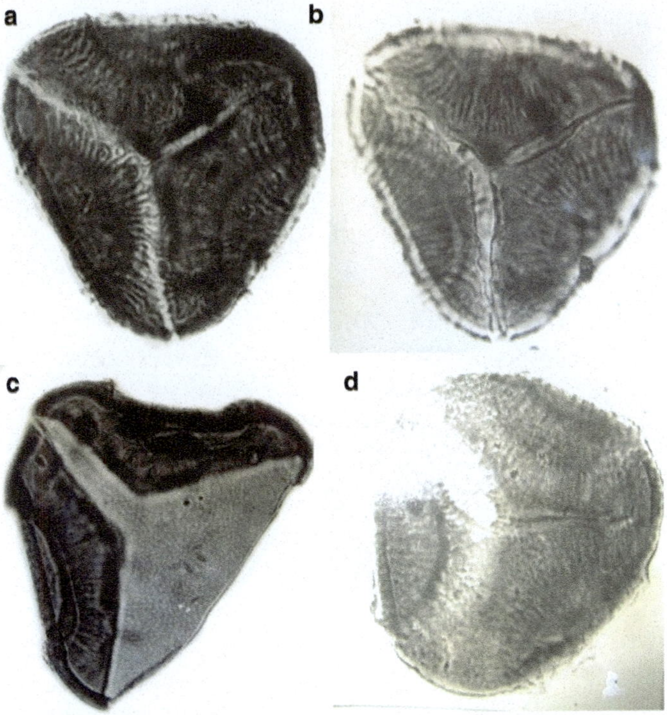

Fig. 2.21 **a** Distal view. **b** Proximal view. **c** Section cut-off showing very fine intra-punctate structure. **d** Full distal view showing areas where 'blisters' are attached

Genus ***Dulhuntyispora*** (Potonie 1956).
Diagnosis: Trilete, cavate. Outline triangular to sub-triangular. Exine consists of two layers; a thick, solid intra-punctate layer overlain by a hyaline layer forming three blisters over the inter-apical area.
Known distribution: Australia Upper Permian (Dulhunty 1946; Balme 1952; Balme and Hennely 1955), Lower Permian of Queensland (Evans 1969).
Discussion: *Dulhuntyispora* was described under the name *Tholosporites* by Balme and Hennelly (1956). However, their publication was preceded by a couple of weeks by Potonie's publication naming the spore *Dulhuntyispora*. Potonie described the spore from Dulhunty's drawing and misinterpreted its construction. Therefore, although the name given to it by Potonie is retained having priority over *Tholosporite*s, Balme and Hennelly's (1956) illustrations must serve as type species.

Dulhuntyispora dulhunty (Potonie 1956) (Fig. 2.21).
Occurrence: Rare. Appears only in Upper Permian samples of Collie–Muja Basin bore B/224′-227′ (Fig. 2.21).

Description: Trilete, cavate. Amb triangular to sub-triangular. Laesurae long, simple, extending to equatorial margin of spore. Exine complex; composed of a thin, smooth ektexine and a thick, two-layered endexine/nexine. Endexine up to 5 μ thick and structured. Structure very fine intra-punctate. Outer ektexine thin, hyaline, tightly fitting over apical area, and expanding into blisters over inter-apical areas. The hyaline ektexine has a pseudo-regulate sculpture formed from its crinkles.
Dimensions: 90–120 μ (10 specimens measured).
Remarks: *Dulhuntyispora* has been described from Collie Basin by Backhouse (1991). Backhouse noted the stratigraphic significance of this spore. It is confined to the Upper Permian.

Supra Sub-turma: Acavatitriletes (Dettmann 1963).
Sub-turma: Azonotriletes (Dettmann 1963).
Genus ***Punctatisporites*** (Potonie and Kremp 1955).
Diagnosis: Trilete. Amb circular or broadly rounded-triangular. Laesurae long. Exine more than 1 μ thick, usually smooth, devoid of sculpture.
Discussion: *Punctatisporites* was emended first by Schopf, Wilson and Bentall (1944) giving a detailed description of the spore. However, according to Schopf et al. *Punctatisporites* still included sculptured trilete spores. Potonie and Kremp emended the genus, restricting it to circular or broadly rounded-triangular forms with laesurae longer than in *Calamospora* and a smooth exine.

Punctatisporites gretensis (Balme and Hennelly 1955) (Fig. 2.22).
Occurrence: Abundant in, and restricted to Sakmarian samples from Collie–Muja Basin (i.e. Bore B/2436-2717′).
Description: Trilete. Amb circular to rounded-triangular. Laesurae simple, extending about three-quarters radius of spore. Exine smooth, 5–7 μ thick.
Discussion: *P. gretensis* is distinguished by its smooth and thick exine which is often pitted and corroded. The corrosion patterns may be misleading and often misinterpreted as structure/sculpture, as the spore corrosion usually starts on the underside of the exine. The range in size is significant and may tempt palynologists to separate the genus into several species. However, in the present study, 100 specimens were studied and 80 measured.
Gradations in size were encountered all along. There was no clear cut differentiation size-wise.
Dimensions: 45–150 μ in diameter (80 specimens measured).
Remarks: *P. gretensis* was recorded and described by Backhouse (1991) from Collie. Backhouse separated *P. gretensis* into two species. He noted the stratigraphic significance of this spore. *P. gretensis* was also described from the Permian of Brazil by Cazzulo-Klepzig et al. (2009).

Punctatisporites sp. (Rigby and Hekel 1977); synonym: *Lalmatiasporites indicus* (D'Rozario and Banerjee) (Fig. 2.23).

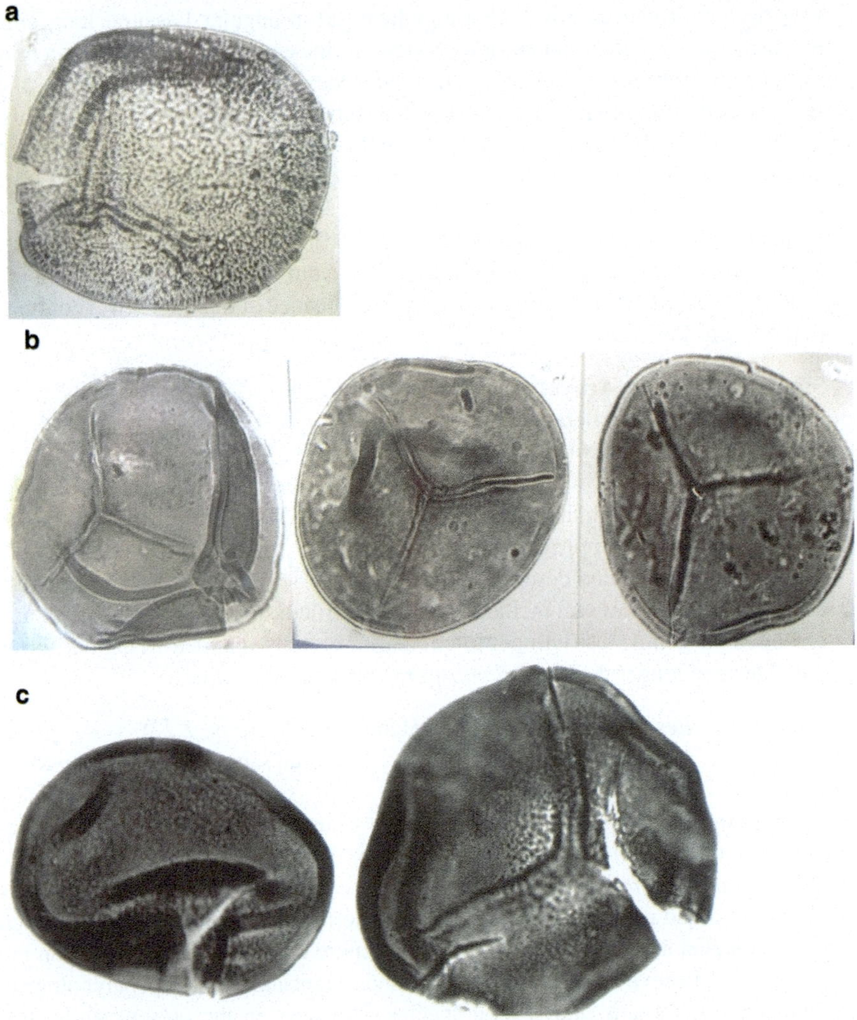

Fig. 2.22 **a** Corrosion of endexine in *P. gretensis*. **b** Thinner and less rigid endexine in immature spore does not lend itself to corrosion. **c** Semi-distal and proximal views. **d** Proximal view showing simple laesurae and very thick exine. **e** Microtome sections of *P. gretensis* showing thick smooth exine. **f**, **g** Tetrads of immature *Punctatisporites gretensis* pollen

Diagnosis: Amb circular to sub-circular. Laesurae complex. Tetrad rays followed by distinct margo from flattened thickened exine. Sculpture consists of low-relief, wide-based verrucae confined to contact area.
Occurrence: Upper Permian, common in Collie–Muja Basin; Bore B/224′-227′, D/201′-212′ and 1/50′.

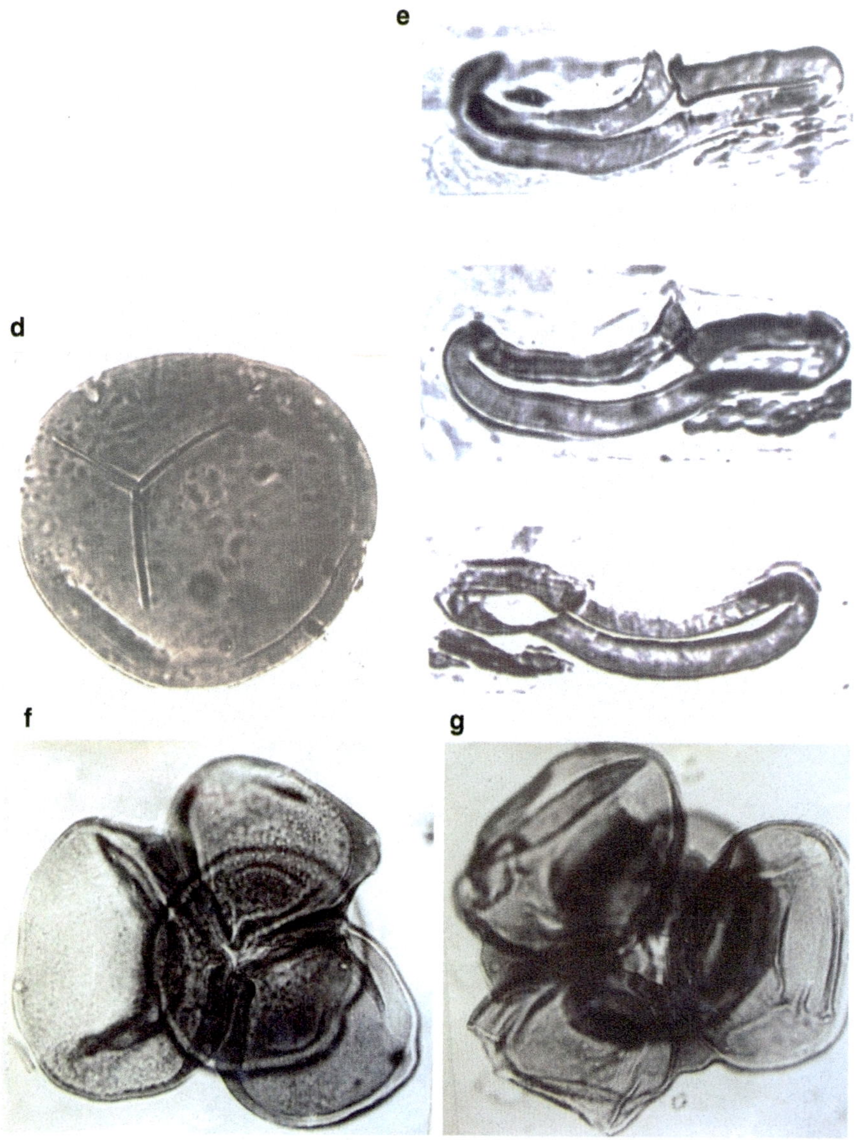

Fig. 2.22 (continued)

Description: Amb circular to sub-circular, or broadly rounded-triangular. Trilete; laesurae complex, extending about three quarters spore radius. Tetrad rays followed by a distinct and characteristic margo. Margo is the result of originally raised commissures (Fig. 2.23e) expressed as folds in the compressed specimens

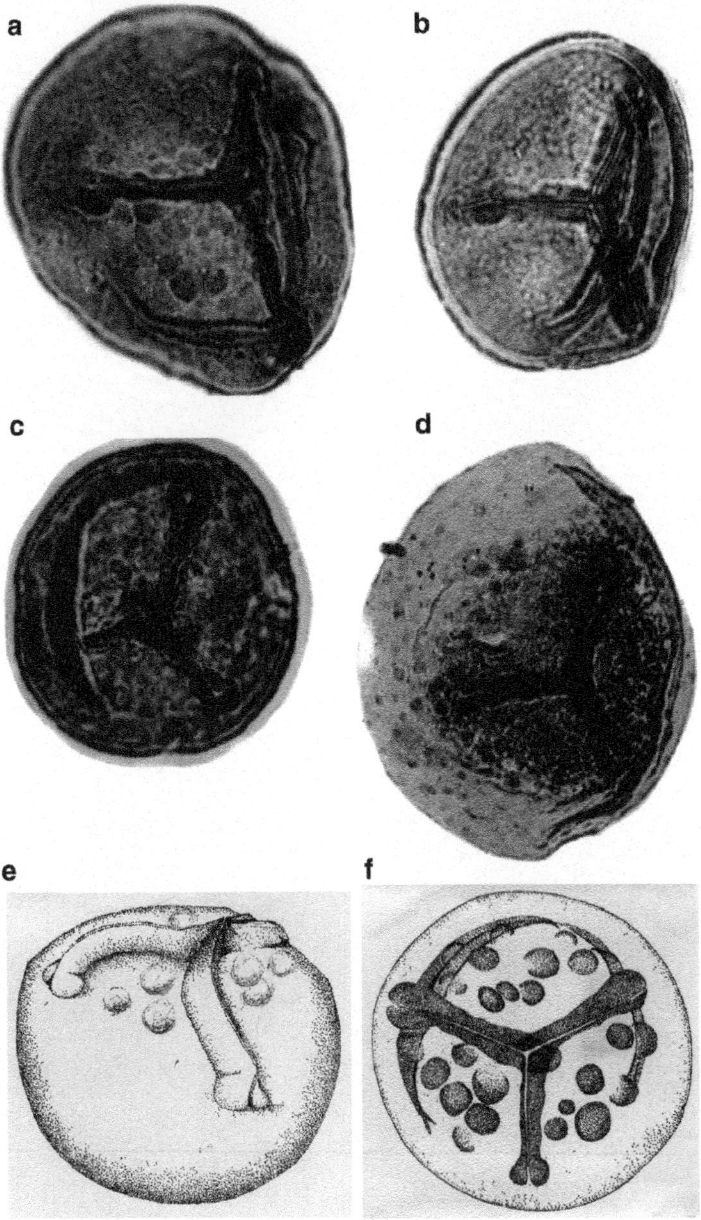

Fig. 2.23 **a–f** Reconstructed non-compressed specimen and compressed pollen

(Fig. 2.23f). Exine smooth to intra-punctate, with low-relief verrucae over contact area.
Dimensions: 45–63 μ (35 specimens measured).
Remarks: *Lalmatiasporites indicus* has been described from the Lower Permian of India. The close resemblance of *L. indicus* to *Punctatisporites gretensis* is apparent and has also been discussed by Banerjee and D'Rozario (1990). However, as also pointed out by these authors, the coarse verrucate sculpture in the proximal area is too distinct to include it in *P. gretensis*.
Rigby and Hekel (1977) designated a pollen of the same appearance as *Punctatisporites* sp. It is treated in the present study as a separate species and I suggest that it be redefined and designated as new species; suggested names, *Punctatisporites lalmatiasporites or Punctatisporites indicus.*

Punctatisporites fungosus (Balme 1963) (Fig. 2.24).
Occurrence: Upper Permian; found only in the uppermost samples of the Upper Permian sequence in the Collie–Muja Basin.
Description: Amb. triangular sub-circular. Trilete; laesurae simple, extending 2/3 to entire spore radius. Spore is large (average 100 μ in equatorial diameter, with a thick exine, 5–7 μ. Exine consists of a very thick endexine (nexine) and very thin sculptured ektexine. The ektexine is usually difficult to distinguish as a separate entity; clearly distinguishable in pollen microtome section (Fig. 2.24c). The endexine is often corroded and has a pitted look. The perforations are randomly distributed and as seen in the microtome section are not a true structure of the exine.
Dimensions: 75–156 μ diameter (20 specimens measured).
Remarks and comparison: *Punctatisporites fungosus* was described by Balme (1963) from the Triassic of Western Australia. Balme does not mention the presence of a thin ektexine. However, this feature is distinguishable only in microtome sections of the spore.
The importance of this species is as a stratigraphic marker in the Collie Basin. The *Punctatisporites fungosus* spore belongs to a plant that appeared at the end of the Permian and continued into the Triassic. Thick exine spores are found today among the ferns of the Dicksoniaceae.

Genus ***Retusotriletes*** (Naumova 1953; Richardson 1965).
Diagnosis: Trilete. Laesurae clear. Amb sub-circular to circular. Contact area has often thickened exine. Exine smooth.
Known distribution: Devonian of Scotland (Richardson 1965). In Australia: Devonian of Carnarvon Basin (Balme and Hassell 1962), Carboniferous of Western Australia (Balme 1960), Upper Devonian of the Canning Basin (Balme and Hassel), Permian of New South Wales (Balme and Hennelly 1956), Permian of Northern Territory (Evans 1964), Devonian of Queensland (de Jersey 1962). In India: Upper Permian of Raniganj coal fields (Bharadwaj 1962), Permian of Brazil (Kazzulo-Klepzig et al. 2009).

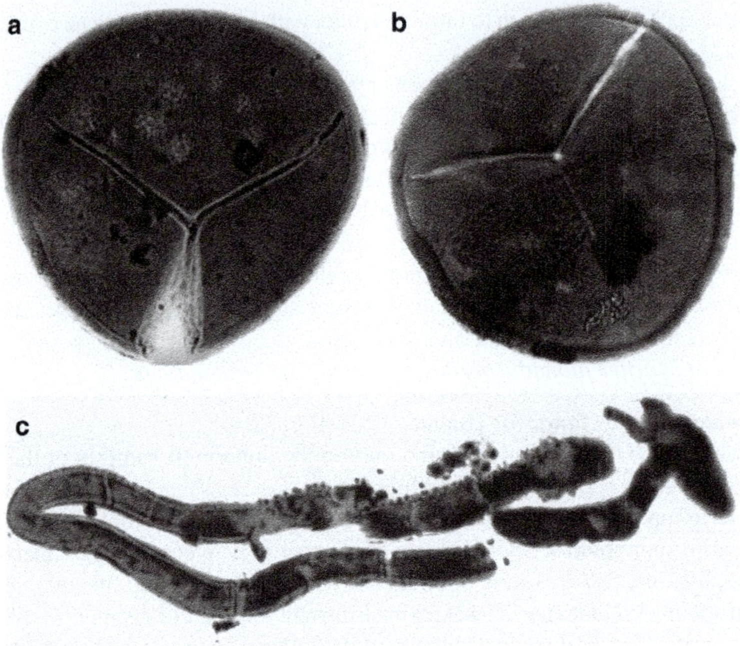

Fig. 2.24 **a** Partly split spore, showing fine, hyaline ektexine. **b** Proximal view, simple laesurae. **c** Microtome section showing the extremely thin gossamer-like structured ektexine breaking up and separating from thick and solid endexine

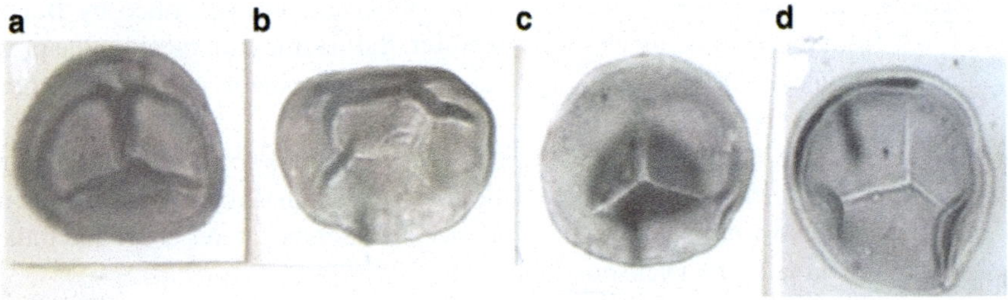

Fig. 2.25 **a** Proximal face, thickenings of exine between laesurae. **b** Slight thickening and folding of exine between proximal mark, **c** thickening around equatorial area, **d** folds in exine

Discussion: *Retusotriletes* was emended by Richardson to include smooth exine forms only.

Retusotriletes diversiformis (Balme and Hennelly 1956; Balme and Playford 1967) (Fig. 2.25).
Occurrence: Sakmarian to Upper Permian of Collie–Muja Basin, Sue No.1 Bore. More common in Upper Permian. Rare in Lower Permian.

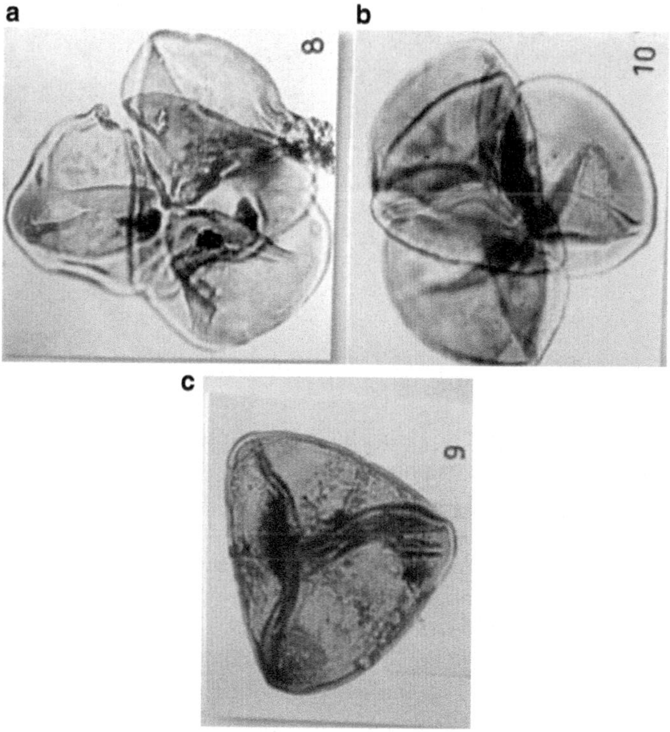

Fig. 2.26 **a** and **b** tetrads. **c** Proximal view with folds along laesurae

Description: Trilete. Amb circular to rounded-triangular. Laesurae distinct, varying in length from short, extending about half way to equatorial margin to almost entire spore radius. Contact area distinct by a thickened exine, often bordered by a curvature which can be partial, rarely perfect. Exine smooth.
Dimensions: 24–36 μ (20 specimens measured).
Remarks: Recorded from Collie Basin by Backhouse (1991).

Genus ***Leiotriletes*** (Naumova) (Potonie and Kremp 1954).
Diagnosis: Trilete. Amb triangular. Exine smooth to punctate. Known distribution: Devonian to Upper Permian in Northern and Southern hemispheres.
Discussion: *Leiotriletes* was used to describe Palaeozoic trilete, triangular spores with a more or less smooth exine. Spores with similar diagnostic features from Mesozoic sediments are described under *Cyathidites Couper*. Ferns of the Dicksoniaceae resemble Leiotriletes.
Known distribution: Permian of Brazil (Kazzulo-Klepzig et al. 2009), Permianof India (Mishra and Jha 2017).

Leiotriletes directus (Balme and Hennelly 1956) (Fig. 2.26).

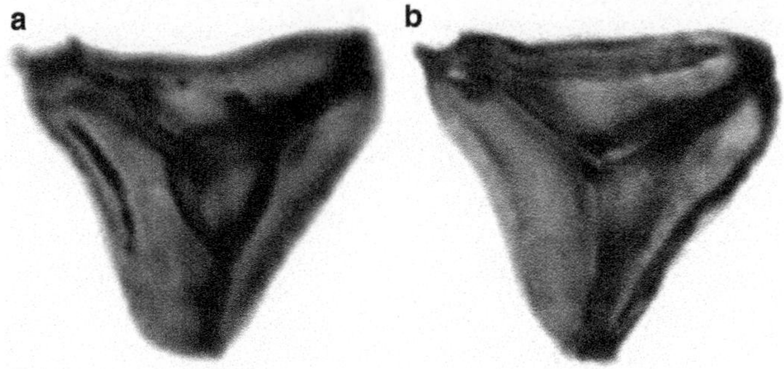

Fig. 2.27 **a** Distal face with ring-like thickening. **b** Proximal face

Occurrence: Widespread in Upper Permian Collie–Muja Basin. Ranges from Artinskian in the samples examined.
Description: Trilete. Amb triangular, sides often slightly convex. Laesurae clearly outlined, extending to or almost equatorial margin, often sinuous and accompanied by folds which indicates raised commissures in the unflattened spores. Exine 1 μ thick or less, smooth or intra-punctate.
Dimensions: 40–50 μ (20 specimens measured).
Remarks and discussion: Spores closely resembling *L. directus* were encountered by Remy and Remy (1957) in the sporangia of *Oligocarpia guthieri*. Backhouse (1991) recorded and described *L. directus* from Collie Basin.

Genus ***Indospora*** (Bharadwaj 1962).
Diagnosis: Amb triangular. Laesurae long, extending to equator. Exine thin, verrucate or bacculate. Three ridges arising from the sub-equatorial region on the proximal face and extending from the apices over equator to meet over the distal polar region, either at a point, or by forming one or more circular or polygonal loops.
Known distribution: Upper Permian of India (Bharadwaj 1962).
Remarks: *Indospora* is distinguished by the unique triradiated ridges on the distal face.

Indospora clara (Bharadwaj 1962) (Fig. 2.27).
Occurrence: Rare, Upper Permian.
Description: Trilete. Amb triangular, sides straight to slightly concave. Laesurae long, extending to equatorial margins. Exine thin, less than 1 μ thick, smooth. Three distal ridges extend from apices, one from each apex to the distal pole area. They meet at a triangular rounded area on the distal face (Fig. 2.25a).
Dimensions: 30–40 μ (4 specimens measured).
Remarks: *I. clara* was described from Collie Basin by Backhouse (1991).

Genus ***Granulatisporites*** (Potonie and Kremp 1954).
Diagnosis: Trilete. Amb triangular. Exine sculpture granulate; grana about 1 μ or less in basal diameter.

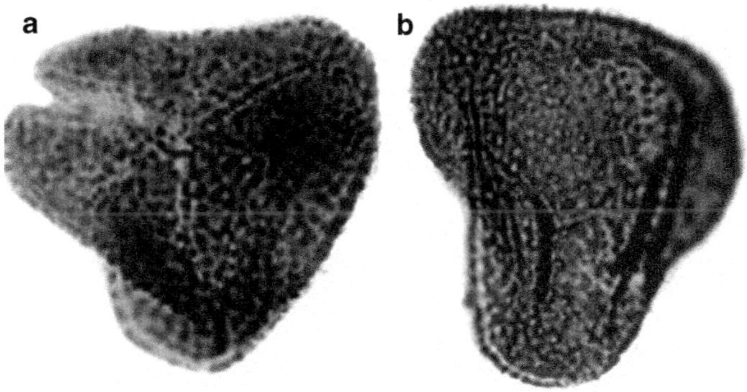

Fig. 2.28 *Indospora clara*; (a) dehiscence of trilete aperture; (b) complete spore

Known distribution: Devonian to Triassic of Northern and Southern hemispheres.
Remarks: *Granulatisporites* is distinguished from *Leiotriletes* by its sculptured exine.
Affinity: Filicopsida.

Granulatisporites cf. granulatus (Potonie and Kremp 1954) (Fig. 2.28).
Occurrence: Upper Permian of Collie Basin.
Description: Trilete. Amb triangular, sides straight to slightly concave, apices broad, rounded. Laesurae extending three quarters or more to equator. Sculpture granulate; grana closely spaced.
Dimensions: 38–45 μ (10 specimens measured).
Remarks: *G. granulatus* in Collie is slightly larger than the forms described by Potonie and Kremp. However, a small difference in size alone does not justify a separate species. The sculpture, however, in the Collie spores is coarser and more densely spaced. Resembles *Cyclogranisporites* sp. described by Backhouse (1991) from the Collie Basin. The sculpture in *Granulatisporites cf. granulatus* is coarser.

Genus ***Acanthotriletes*** (Naumova) (Potonie and Kremp 1955).
Diagnosis: Trilete. Amb triangular. Sculpture of exine echinate; spines sharp-edged.
Known distribution: Lower Carboniferous to Triassic of Northern and Southern hemispheres.
Remarks: *Acanthotriletes* was first reported and described from the Carboniferous of Russia (Naumova 1961). The genus was later validated by Potonie and Kremp by naming a type species, choosing a form described by Knox (1950) under the name *Spinososporites*. The latter was dropped by Potonie and Kremp because it embraced monolete as well as trilete forms. *Acanthotriletes is* distinguished from *Apiculatisporites* mainly by its triangular amb compared to circular to sub-circular outline in *Apiculatisporites*.

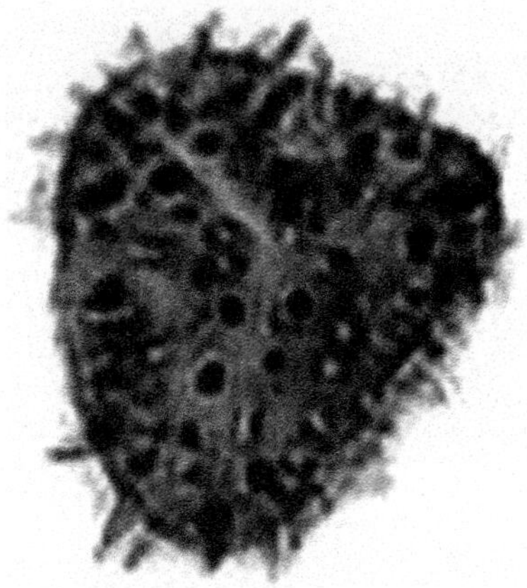

Fig. 2.29 Proximal view

Acanthotriletes tereteangulatus (Balme and Hennelly 1956) (Fig. 2.29).
Occurrence: Common in Artinskian to Upper Permian of Collie–Muja Basin.
Description: Trilete. Amb triangular; sides often slightly concave, apices rounded. Laesurae simple, extending to equatorial margin. Sculpture echinate; spines wider at their base and thinning towards their sharp ends; vary in size in same specimen, from 1 to 4 μ in length, and about 1 μ wide at their base.
Dimensions: 25–40 μ (15 specimens measured).
Remarks and Comparison: Main distinctive features are the straight to concave sides and relatively long spines.

Genus ***Apiculatisporites*** (Ibrahim 1933; Potonie and Kremp 1954).
Diagnosis: Trilete. Amb circular to sub-circular. Sculpture of exine echinate; elements are wide-based cones.
Known distribution: Devonian to Permian of Northern and Southern hemispheres.
Discussion: *Apiculatisporites* is distinguished from *Anapiculatisporites* in the absence of sculptural elements on the proximal face in *Anapiculatisporites.*

Apiculatisporites cornutus (Balme and Hennelly 1956) (Fig. 2.30).
Occurrence: Upper Permian of Collie–Muja Basin.
Description: Amb sub-circular to rounded-triangular. Trilete; laesurae not clearly outlined, extending about three-quarters of spore radius. Sculpture echinate; cones are wide-based and sharp-tipped, about 4–5 μ in diameter at their base and up to 3.5 μ long.
Dimensions: 24–36 μ (12 specimens measured).
Remarks: *A. cornutus* is distinguished by its heavy cones and indistinct laesurae. The species has been described from Collie Basin by Backhouse (1991) under *Brevitriletes cornutus* Balme and Hennelly. cf. *Horriditriletes unicus* described by Tiwari (1964) from the Permian of India closely resembles *A. cornutus* and may prove to be its synonym on closer examination.

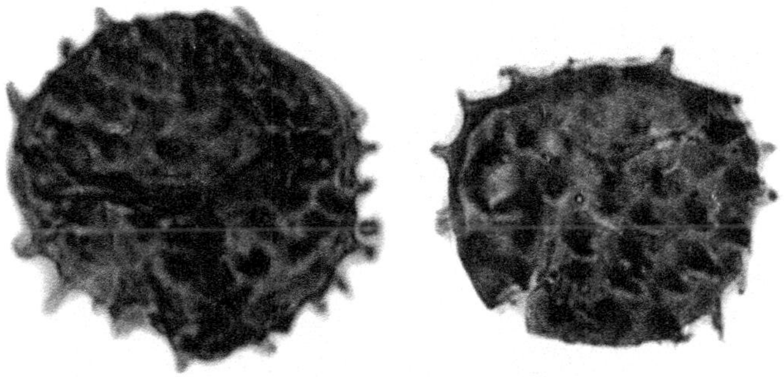

Fig. 2.30 *Apiculatisporites cornutus*

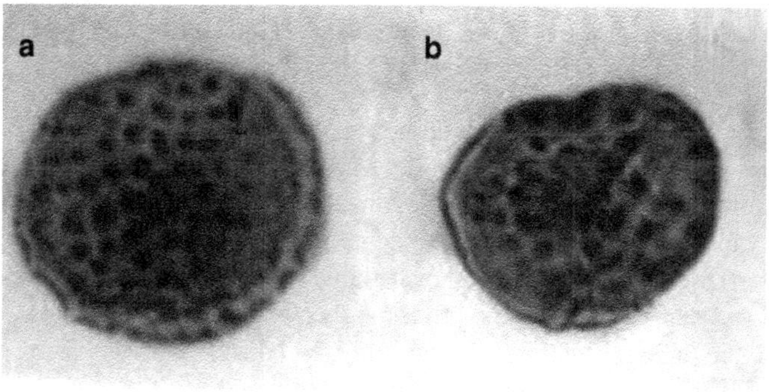

Fig. 2.31 **a** Distal face, **b** proximal face

Apiculatisporites levis (Balme and Hennelly 1956) (Fig. 2.31).
Occurrence: Lower to Upper Permian of Collie–Muja Basins. More common in Sakmarian than Upper Permian of Collie Basin (Bore B-2513′).
Description: Exceptionally small spore, rarely exceeding 28 μ in equatorial diameter. Amb circular. Laesurae distinct, simple, extending 2/3 to 3/4 of spore radius. Exine sculptured with tiny spines.
Dimensions: 21–27 μ (15 specimens measured).
Remarks: *A. levis* is distinct from *A. cornutus* mainly by its minute size. Backhouse (1991) described a similar spore from Collie Basin under *Rattiganispora* (Fig. 2.31).

Genus ***Anapiculatisporites*** (Potonie and Kremp 1955).
Diagnosis: Trilete. Amb triangular to rounded-triangular. Sculpture reduced on proximal face to smooth over the contact area. Sculpture is echinate consisting of wide elements; conical at their base and spinate at their ends.
Known distribution: Upper Carboniferous of the Ruhr (Potonie and Kremp 1955), Upper Carboniferous of Canada (Kosanke 1950). Permian of New South Wales (Balme and Hennelly 1956), Permian of South Africa (Hart 1963).

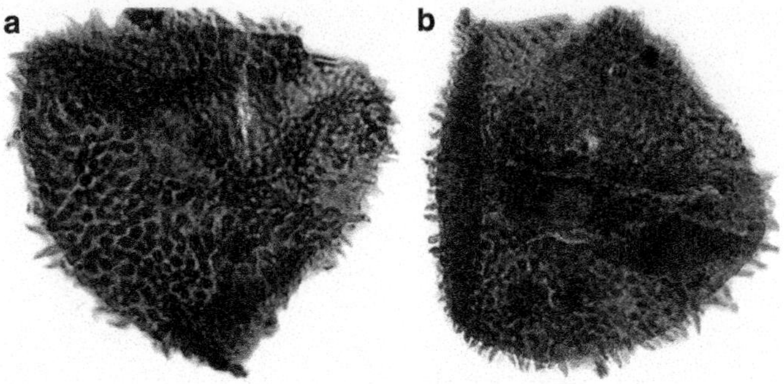

Fig. 2.32 **a** Side view, long projections visible, **b** semi-proximal view, laesurae visible

Remarks: *Anapiculatisporites* is distinguished from *Apiculatisporites* by smooth proximal face in the former.

Anapiculatisporites ericianus (Balme and Hennelly 1956) (Fig. 2.32).
Occurrence: Upper Permian of Collie–Muja Basin, Sue No.1 bore hole.
Description: Amb triangular. Sides convex. Trilete; laesurae crested, long, not clearly outlined, extending almost to equatorial margin of spore. Proximal face of low relief, pyramidal, while distal face highly convex. Sculpture echinate, reduced in size and density towards the proximal face and absent altogether over contact area. Exine of equal thickness on proximal and distal faces. Spine are prominent and are more or less uniform in thickness and length; 2–3 μ wide at their bases, closely spaced and up to 6 μ long.
Dimensions: 45–80 μ (20 specimens measured).

Genus ***Microbaculispora*** (Bharadwaj 1962).
Diagnosis: Amb. triangular to triangular-rounded. Trilete mark clearly outlined; laesurae long and crested, expressed in exinal folds accompanying rays. Proximal face pyramidal and thin-walled. Contact area smooth. Distal face highly convex, thick-walled and sculptured. Sculpture granulate, echinate or echino-granulate. Distribution of sculptural elements very regular over spore exine, excluding contact area on proximal face, where sculptural elements decrease in size until fading over contact area.
Known distribution: Permian of the Southern hemisphere (Balme and Hennelly 1956; Hoeg and Bose 1960; Bharadwaj 1962; Hart 1963; Evans 1964; Tiwari 1965).
Discussions: *Microbaculispora* was proposed by Bharadwaj (1962) for bacculate trilete spores whose sculptural elements decrease in size on proximal face and are practically absent from contact area. The genus is here emended to emphasise the crested laesurae, pyramidal proximal face with a thin unsculptured exine, and a thicker sculptured exine on the distal face, as characteristic features of the genus. The nature of the sculptural elements is left to specific diagnosis.

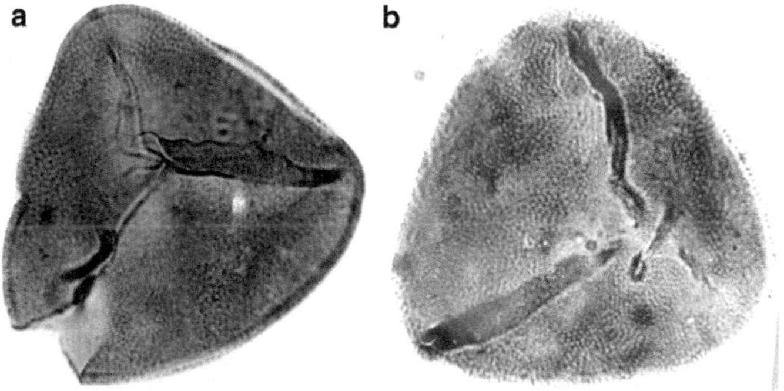

Fig. 2.33 **a** Fine micr族ranulate structure, notice detached ektexine. **b** Folds following laesurae

Anapiculatisporites lacks crested laesurae and thinning of proximal exine characteristic of Microbaculispora. Furthermore, a very regular sculpture characterises *Microbaculispora.*

Microbaculispora trisina (Balme and Hennelly 1956) New Comb (Fig. 2.33).
Occurrence: Upper Permian of Collie–Muja Basin. Sue No.1 Bore.
Description: Amb triangular, convex sides. Trilete; laesurse clearly outlined, extending to equatorial margin of spore. Undulating folds following tetrad rays. Proximal face pyramidal. Distal face convex. Exine about 2 μ thick on distal face, thinning out towards proximal face. Sculpture granulate; grana vary in height from ≤1 to 1 μ, closely packed forming a regular pattern in surface view, flattening towards proximal face, absent from contact area.
Dimensions: 50–80 μ (20 specimens measured).
Remarks and comparisons: *M. trisina* is distinguished mainly by the size of its sculptural elements, which are grana not exceeding 1 μ in diameter and height. *M. trisina* reported and described by Backhouse (1991) from Collie Basin, and from the Karroo Basin by Anderson (1977).

Microbaculispora micronodosa (Balme and Hennelly 1956) (Fig. 2.34).
Occurrence: Upper Permian of Collie–Muja Basins, Sue No.1 bore hole.
Description: Trilete. Sinuous folds accompanying tetrad scar, signifying ridges before compression. Amb triangular with convex sides. Proximal face pyramidal and devoid of sculpture. Distal face convex and sculptures: Sculpture of exine granulate; grana about 1 μ in diameter and 1–2 μ in height, diminishing in size on the proximal face and absent altogether over contact area.
Dimensions: 36–57 μ (15 specimens measured).
Remarks and comparisons: *M. micronodosus* is distinguished from *M. trisina* by possessing larger sculptural elements relative to its overall size.

Microbaculispora villosa (Balme and Hennelly 1956; Bharadwaj 1962) (Fig. 2.35).
Occurrence: Upper Permian of Collie–Muja Basins. Sue No.1 bore hole.

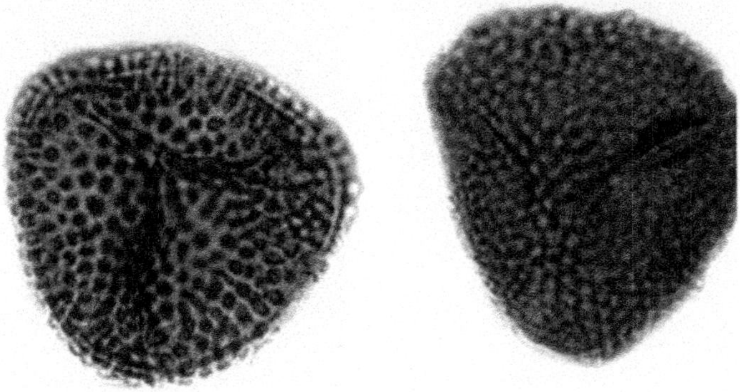

Fig. 2.34 Two focal levels of *Microbaculispora micronodosa*

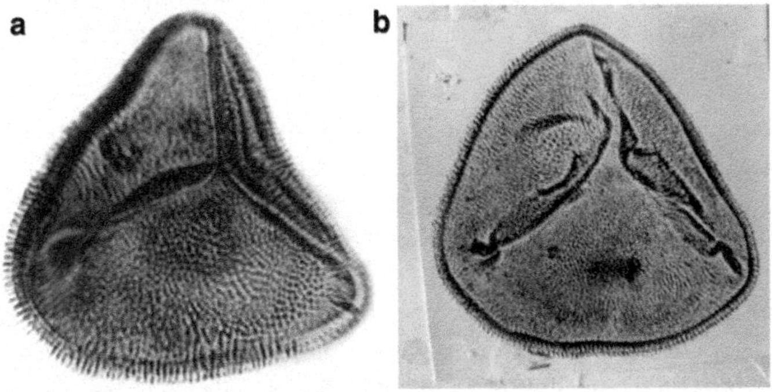

Fig. 2.35 a Laesurae accompanied by folds. b Coarse micro-reticulate pattern made by bases of projections. b Prominent sculpture seen in optical section

Description: Trilete. Amb triangular; sides convex. Laesurae long, extending to equatorial margin, followed by broad sinuous folds. Exine thick on distal face, thinning towards proximal face; contact area thin and devoid of sculpture. Sculpture consists of densely spaced rod-like elements. These are uniform in shape and size, forming a perfect mini-reticulate pattern in surface view (Fig. 2.35). Sculptural elements up to 4 μ in height.
Dimensions: 54–90 μ (50 specimens measured). *Microbaculispora villosa* is distinguished from *M. trisina* mainly by the length of its processes which in the former exceed 1.5 μ.

Microbaculispora tentula (Tiwari 1965) (Fig. 2.36).
Occurrence: Lower Permian (Sakmarian) of Collie-Muja Basin, Sue No. 1 bore.
Description: Trilete. Amb triangular, sides straight to slightly concave. Laesurae long, extending to equatorial margin of spore. Broad sinuous folds accompanying

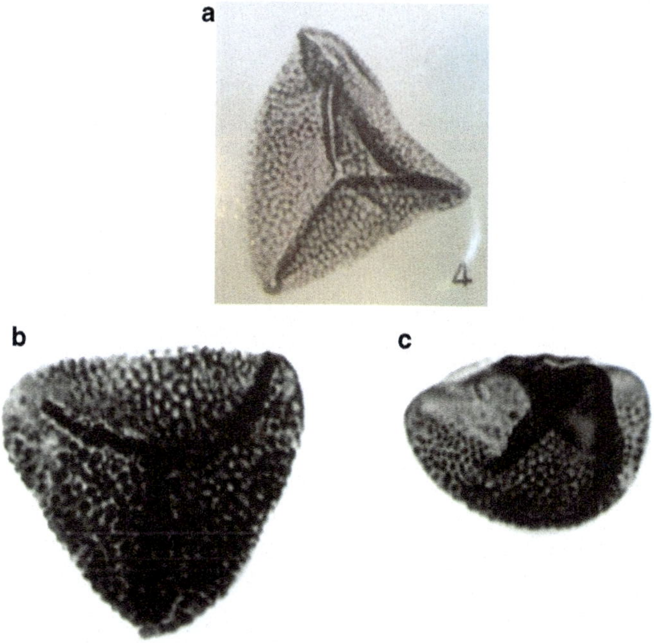

Fig. 2.36 **a** Laesurae accompanied by folds, seen from distal face through to proximal. **b** Side view showing half distal and half proximal faces. **c** Proximal face; laesurae accompanied by folds

commissures. Proximal face rounded pyramidal with thin exine. Sculpture grano-verrucate; elements of low relief, 1–3 μ in diameter, evenly shaped and spaced. Sculpture absent on proximal face.
Dimensions: 40–66 μ (30 specimens measured).
Remarks and comparisons: *M. tentula* is distinguished from the other species of *Microbaculispora* described in this study mainly by its sculpture which differs from the other species of the same genus. Of significance is its distribution; *M. tentula* is restricted to the Lower Permian (Sakmarian), whereas *M. trisina*, *M. micronodosa* and *M. villosa* are all found in the Upper Permian of Collie Basin. The stratigraphically restricted distribution is significant and justifies the species separation.

Pseudoreticulatispora pseudoreticulata (Balme and Hennelly 1956) (Fig. 2.37).
Occurrence: Lower Permian of Collie–Muja Basin. Frequent in Bore site B/1975, 1945′; Bore Site D/1938′, 1936′.
Description: Amb triangular-rounded. Trilete; laesurae distinctive, extending almost to equatorial margin of spore, followed by folds. Proximal face pyramidal. Exine on distal face up to 3 μ thick, thinning towards proximal face. Sculpture verrucate. Verrucae of low relief, regular in shape, 2–3 μ in diameter, closely spaced, often coalescing, diminishing in elevation towards proximal face and absent on contact area.

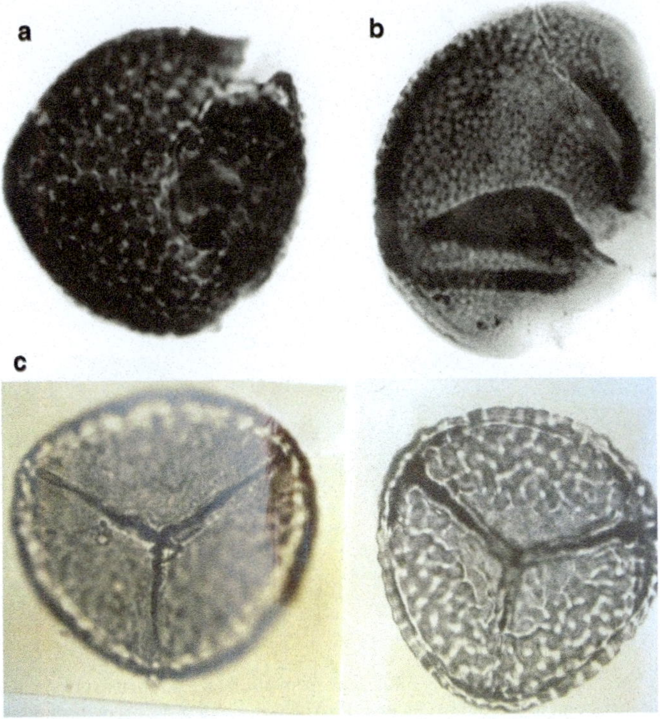

Fig. 2.37 **a** Distal face with part of laesurae seen through. **b** Side view part distal heavily sculptures and part proximal faint sculpture

Dimensions: 60–80 μ (18 specimens measured).
Remarks: *P. pseudoreticulata* is distinguished by its perfect verrucate sculpture. The species has been recorded from Collie Basin by Backhouse (1991). There is a close resemblance to the living fern *Phylloglossum drummondii* (Fig. 2.36c).

Genus ***Balmeospora*** (Backhouse 1988) (Fig. 2.37).
Balmeospora gliksoniae (Backhouse 1988) (Fig. 2.38a, b).
Occurrence: Upper Permian of Collie–Muja Basins, Bore Site B/224-227′; Bore Site D/207-212′, Muja 1/50′.
Description: Trilete. Amb circular. Laesurae simple, extending 3/4 to equatorial margin. Endexine 3–4 μ thick and smooth. Ektexine hyaline and sculptured. Sculpture pseudo-reticulate consisting of irregularly shaped processes with rounded ends, closely spaced and often fused at their tips.
Dimensions: 60–150 μ; mean 90 μ (40 specimens measured).
Remarks: *Balmeospora gliksoniae* has been designated by Backhouse (1988). The species has also been recorded from the Collie Basin by Backhouse (1991). *B. gliksonaiae* closely resembles the spore of the fern *Todea sp.* of the family Osmundaceae. Side by side they appear identical (Fig. 2.38c—*Todea*).

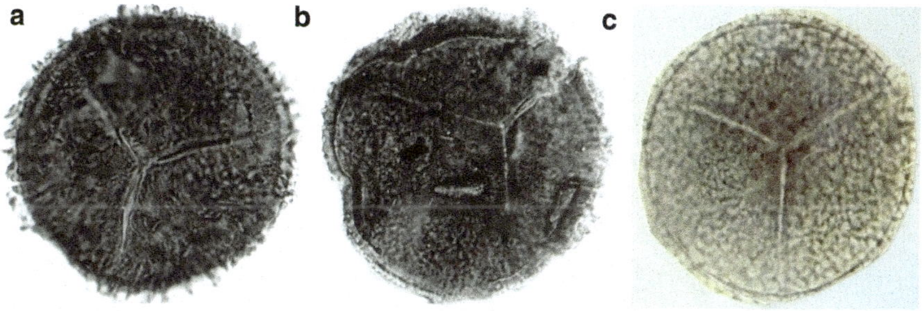

Fig. 2.38 **a** Coarse sculpture, pseudo-echinate. **b** Semi-proximal view, thick nexine and dense sculpture of ektexine. **c** *Todea sp*

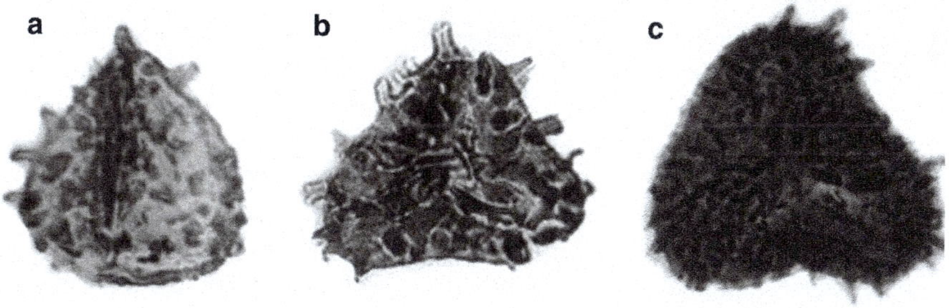

Fig. 2.39 **a** Distal view, **b**, **c** proximal view

Genus ***Neoraistrickia*** (Potonie 1956).

Diagnosis: Trilete. Amb triangular. Sculpture varies in shape from spinate to bacculate to pilate.

Known distribution: Permian of Western Australia (Balme and Hennelly 1956), Permian of Tanganyika (Hart 1960), Upper Permian of Raniganj Coal (Bharadwaj 1962; Salujha 1964), Permian of Brazil (Cazzulo-Klepzig et al. 2009); Permian of India (Mishra and Jha 2017).

Discussion: *Neoraistrickia* is distinguished from *Raistrickia* (Potonie and Kremp 1955) in outline; *Raistrickia* being circular in outline, while *Neoraistrickia* is distinguished by a triangular Amb. *Horriditriletes* (Bharadwaj and Salujha 1967) closely resembles *Neoraistrickia ramosa*, and may well be its synonym. (*Neoraistrickia*) *Horriditriletes ramosus* is recorded from Lower Permian of Godavari Graben, South India (Mishra and Jha 2017).

Neoraistrickia ramosa (Hart 1960) (Fig. 2.39).

Occurrence: Lower to Upper Permian of Collie-Muja Basin.

Description: Trilete. Amb triangular; sides straight to concave, apices rounded. Laesurae extending almost to equatorial margin. Sculpture of exine irregular in shape, size and distribution; bacculate, pilate or echinate.

Dimensions: 40–60 μ (15 specimens measured).

Remarks and comparison: *Neoraistrickia ramosa* is distinguished from the type species by the irregularity of the shape and distribution of sculpture. Species closely resembling *Neoraistrickia ramosa* were described by Bharadwaj and Salujha under the genus *Horriditriletes*. The present study treats *Neoraistrickia* as synonymous with *Horriditriletes*, but retains the name *'Neoraistrickia'*. *Horriditriletes ramosus* was reported from the Permian of Cuhuparbita and Hura Basins in East India (Banerjee and D'Rozario 1990).

Turma: MONOLETES (Ibrahim 1933).
Genus ***Laevigatosporites*** (Ibrahim 1933).
Diagnosis: Amb oval to sub-circular. Monolete; simple laesurae. Exine smooth to intra-punctate.
Known distribution: Upper Carboniferous to Tertiary of Southern and Northern hemisphere.
Discussion: *Laevigatosporites* is distinguished from other monolete spores mainly by its smooth to almost smooth exine, and simple monolete scar.
Laevigatosporites was established by Potonie and Kremp (1954) for monolete more or less smooth, oval to sub-circular spores. Potonie and Kremp designated *Laevigatosporites* latus Kosanke (1950) as type species for *Latosporites*. However, *Laevigatosporites* latus as described by Kosanke fits perfectly into *Laevigatosporites* Ibrahim. The difference noted by Potonie and Kremp is in the outline, being more sub-circular than oval in *Latosporites*, and having less inflated distal face than in *Laevigatosporites*. However, the convexity of the distal face is rarely determinable, since most specimens are compressed in the equatorial plane and are observed in polar view only. The outline of *Laevigatosporites* is not considered by the present author as being a distinct enough feature to justify the separation of the two genera.
Monolete smooth spores resembling *Laevigatosporites* were separated from sporangia of *Bowmanites bifurcates* by Andrews and Mamay (1951) and attributed to the Sphenopsidae. Similar spores were also isolated from sporangia of *Ptychocarpus unitus* (Remy 1957).

Laevigatosporites colliensis (Balme and Hennelly 1956) (Fig. 2.40).
Occurrence: Uncommon. Occurs occasionally in Artinskian to Upper Permian samples from Collie–Muja Basin.
Description: Monolete. Amb oval to sub-circular. Outline smooth. Laesurae simple, extending to almost whole length of spore; usually followed by folds. Exine smooth and thin, less than 1 μ.
Dimensions: Length: 54–69 μ; width: 30–54 μ (15 specimens measured).
Remarks and comparisons: *Laevigatosporites colliensis* is distinguished mainly by its thin, smooth exine. *L. colliensis* has been recorded and illustrated from Collie Basin by Backhouse (1991). Backhouse presented a similar argument as outlined above, namely that *L. colliensis* "*is presented in a variety of compressional positions at all levels within its range and there seems to be no sound reason for regarding transversely folded specimen as a separate species.*"

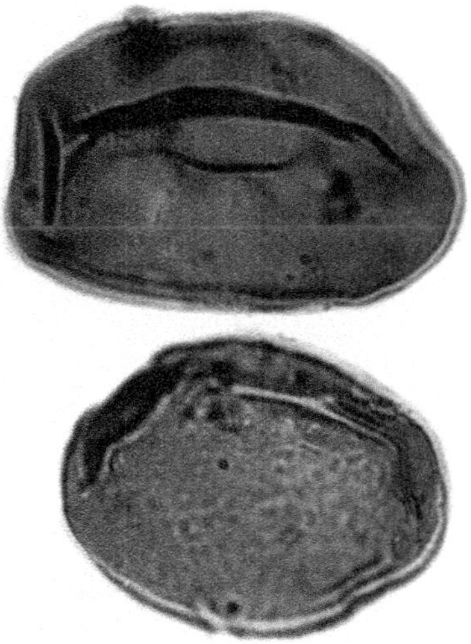

Fig. 2.40 *Laevigatosporites colliensis* showing folds

Anteturma: POLLENITES

Turma: SACCITES (Erdtman 1957).

Sub-turma: Monosaccites (Potonie and Klaus 1954).

The monosaccate pollen are attributed to the ***Cordaitales***, which are an extinct group of Gymnospermous trees known to have reached a height of more than 30 m. They had world wide occurrence during the Permian.

The interesting TEM (transmission electron microscope) study of monosaccate pollen by Zavialova et al. (2004) demonstrates the basic structure of all monosaccate pollen. The relationship between the corpus and saccus is clearly visible in these ultra-sections. The structure of the saccus can be observed in detail as well as its extension within the saccus. This study reviews some of the monosaccate pollen described under various generic names including *Florinites*, *Potoniesporites*, *Nuskoisporites* and others, and suggests the inclusion of them under the genus *Cordaitina*. I suggest that *Barakarites* and *Parasaccites* may also be proven to belong to *Cordaitina.* However, as Zavialova comments, TEM studies of all these pollen types need to be carried out to compare their ultra-structure and sort out the genera on a valid basis. The distribution in time of the different pollen types is also an important factor in the consideration of generic and specific divisions.

Genus ***Densipollenites*** (Bharadwaj 1962).

Diagnosis: Monosaccate. Corpus outline circular to sub-circular or oval. Nexine (endexine) varies in thickness. Large saccus usually folded. Tetrad mark absent. Ektexine detached on one face (distal?) and attached over a small area on the opposite (proximal?) face.

Known distribution: Upper Permian of India, Raniganj coalfield, Bihar (Bharadwaj 1962; Bharadwaj and Salujha 1964). Upper Permian of Antarctica, Amery Formation, Prince Charles Mountains (Balme and Playford 1969).

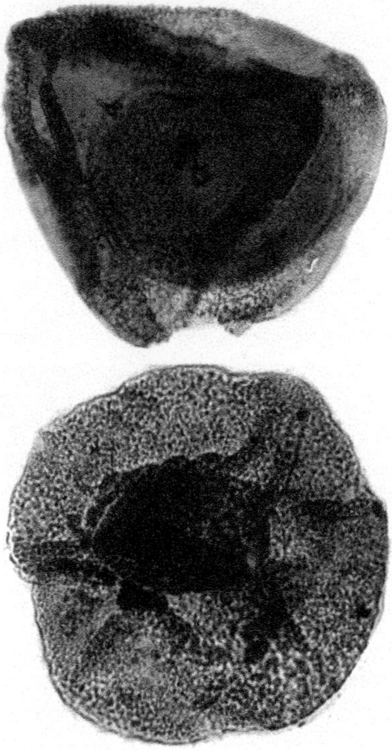

Fig. 2.41 Corpus seems to 'float' within sac

Discussion: Wilson (1960) in his discussion of *Florinites* which resembles *Densipollenites* interpreted it as having the ektexine detached from the entire proximal face and attached only over the central area of the distal face. However, *Florinites* differs *from Densipollenites* in being bilaterally symmetrical and possessing a smaller saccus, reflected by the absence of prominent folds in *Florinites*. Bharadwaj noted the dense structural elements.
The irregular position of the corpus in *Densipollenites* as a result of the ektexine being detached from the nexine (endexine) over the greater part of the corpus, distinguishes this genus from *Potoniesporites*.

Densipollenites indicus (Bharadwaj 1962) (Fig. 2.41).
Occurrence: Found only in Lower Permian of Wilga Basin. Bore 3/80′, 110′, 189′.
Description: Monosaccate. Outline of corpus circular, varying in size. Tetrad mark absent. Ektexine attached to nexine on one face. The rest of nexine is loose and often folded or distorted. Structure of nexine granular, coarse, covering evenly entire area.
Dimensions: 70–150 μ of total pollen grain; corpus diameter 25–105 μ (46 specimens measured).

Remarks and comparison: *D. indicus* is characterised by the distinct variation of the ratio of the corpus to saccus from one specimen to another. This feature may easily lead to separating the species on these grounds. However transitions are evident.

Genus ***Culleisporites*** (Grebe 1957)
Diagnosis: Monosaccate, roughly radially symmetrical. Amb circular, sub-circular or lobed. Outline of corpus circular to sub-circular. Ektexine detached equatorially on distal face.
Nexine thinner than ektexine. Distal saccus lobed forming a notched distal tenuiexinous area. This configuration was stressed by Grebe as a distinct feature identifying this genus. Structure dense, intra-granulate to intra-reticulate with conspicuously long columellae projecting into saccus. No proximal mark was observed by Grebe (1957) in any of the specimens, and she noted that the 'trilete mark' in Leschick's specimen is vague.
Known distribution: Upper Permian of Germany (Leschik 1956; Grebe 1957).

Culleisporites densus (Leschik 1956) (Fig. 2.42).
Occurrence: Upper Permian of Collie–Muja Basin; Bore B/224-227′; Site D/201-212′.
Description: Monosaccate, radially symmetrical. Amb circular, sub-circular or lobed. Corpus outline circular. Nexine, thin-structured, detached equatorially on proximal face and sub-equatorially on distal face. Saccus roots may be slightly or significantly notched. In the case of the latter the pollen appears as having multiple sacci. Structure of attached ektexine intra-granulate to intra-reticulate towards sacci rims. In cross-section the columellae of the reticulum are distinct.
Dimensions: 52–80 μ; corpus diameter 52–80 μ (15 specimens measured).
Discussion: Some living forms of conifers, such as *Araucaria* may have several sacci, although essentially monosaccate (Fig. 2.42a). Likewise bisaccate pollen may be trisaccate, e.g. *Podocarpus*. Variations within one pollen sac are not uncommon and should be taken into account when separating dispersed pollen.

Genus ***Bascanisporites*** (Balme and Hennelly 1956).
Diagnosis: Amb circular, sub-circular or lobed. Tetrad mark absent or faint. Nexine thicker than ektexine. Structure of ektexine fine.
Known distribution: Upper Permian of New South Wales (Balme and Hennelly 1956), Upper Permian of Queensland (Hill and Woods 1964), Upper Permian of Antarctica (Balme and Playford 1967).
Discussion: *Bascanisporites* is distinguished from *Culleisporites* by a thick nexine in *Bascanisporites*, which is responsible for the strong and clear outline of the pollen grain, and the absence of folds in the latter.

Bascanisporites undosus (Balme and Hennelly 1956) (Fig. 2.44a, b).
Occurrence: Artinskian to Upper Permian of Collie–Muja Basin; Bore B/224-227′, B/1641′, B/2149′, D/201-212′. More common in the Upper Permian than in Artinskian samples.

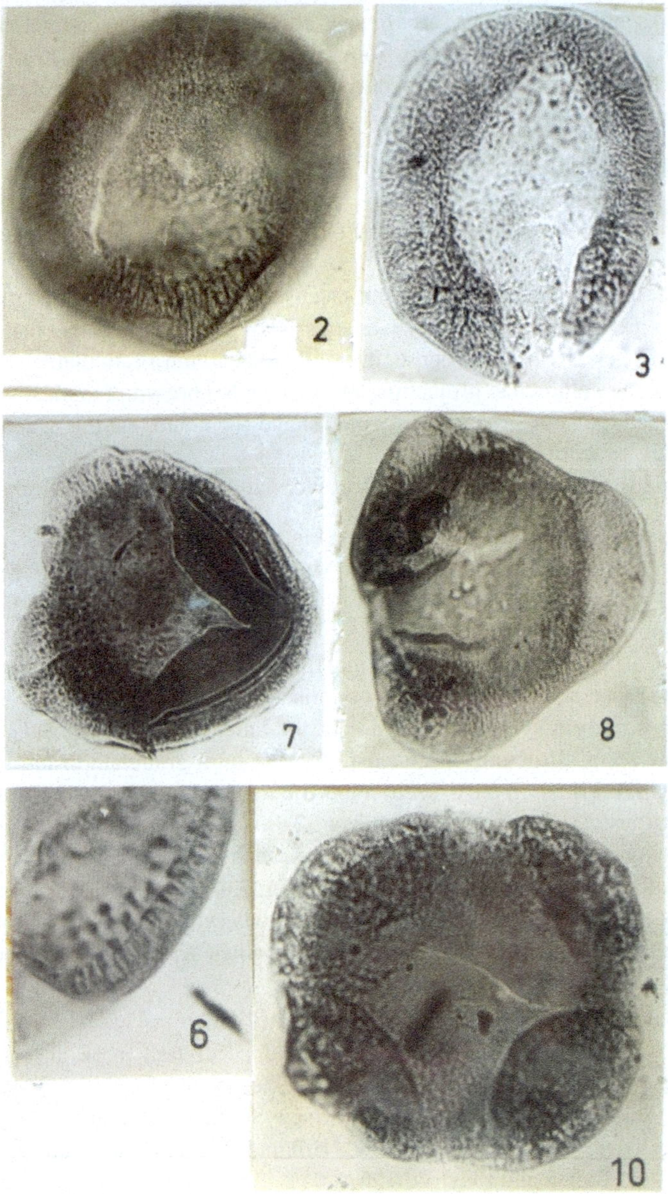

Fig. 2.42 Varies from monosaccate (2, 3) to lobed (8) to tri-saccate (7) to four-saccate (10). (6) detail of structure

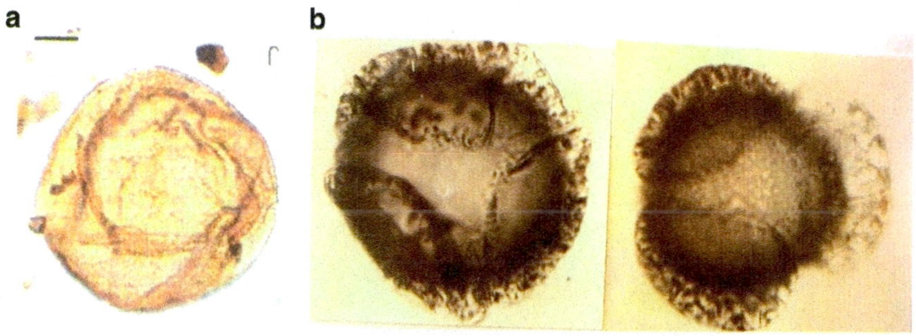

Fig. 2.43 **a** *Araucaria balansae.* **b** *Podocarpus compactus*

Description: Monosaccate. Radially symmetrical. Amb circular to sub-circular. Nexine thicker than ektexine which is the main feature distinguishing it from *Culleisporites.* The structure in *Bascanisporites* sacci is very fine intra-punctate compared to the coarse saccus structure in *Culleisporites.*
Dimensions: 50–100 μ, range of corpus diameter 45–80 μ (18 specimens measured).
Remarks: *Bascanisporites undosus* is recorded and described from the Permian of Collie Basin by Backhouse (1991) (Fig. 2.43).

Genus ***Barakarites*** (Bharadwaj and Tiwari 1964).
Diagnosis: Monosaccate, radially symmetrical. Amb circular to rounded-triangular. Tetradmark clearly visible. Laesurae simple. Ektexine structure fine, intra-punctate.
Known distribution: Permian of New South Wales (Balme and Hennelly 1956), Upper Permian of Raniganj coal field, India (Bharadwaj and Tiwari 1964), Bharadwaj and Salujha 1964. Karroo series in Tanganyika (Hart 1963), Lower Permian of Congo (Hoeg and Bose 1960), South India (Mishra and Jha 2017).
Discussion: *Barakarites* is distinguished from other monosaccate forms by the very faint outline of inner body, the result of an extremely thin endexine/nexine. So much so that Bharadwaj and Tiwari specially noted 'the presence of an inner body' that may not be easily distinguished.

Barakarites cf. rotatus (Bharadwaj and Tiwari 1964) (Fig. 2.45).
Occurrence: Upper Permian in Collie–Muja Basins; Bore B/224-227′, 2125′; Bore D/201-213′.
Description: Monosaccate, radially symmetrical. Amb circular to rounded triangular. Tetrad mark not always distinct. When visible small; laesurae extend only 1/3 way to equator.
Nexine significantly thicker than ektexine. Ektexine detached equatorially on proximal face and sub-equatorially on distal face. Saccus radially folded. Structure fine intra-punctate.
Dimensions: Diameter 92–140 μ, corpus diameter 68–100 μ (10 specimens measured).

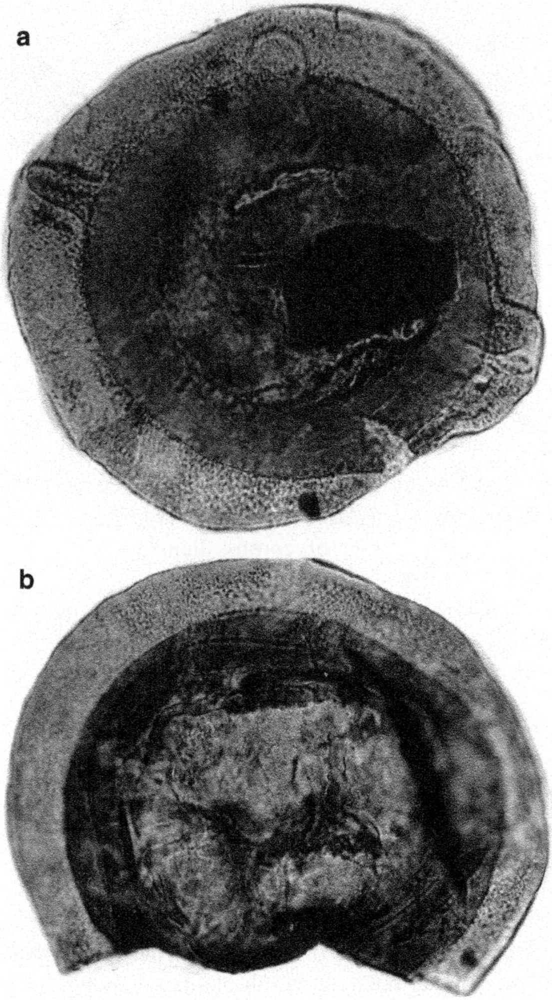

Fig. 2.44 *Bascanisporites undosus;* (a) proximal face; (b) distal face

Remarks and discussion: *Barakarites* cf. *rotatus*. Differs from *Barakarites rotatus* in possessing a thick nexine while *B. rotatus* is characterised by an extremely thin nexine. The thick nexine in *Barakarites* sp. gives the corpus a very dark appearance, while the thin nexine in *B. rotatus* makes the corpus appear hyaline. Backhouse (1991) places forms closely resembling *B. sp*. in *Plicatipollenites sp*. So do also Banerjee and D'Rozario (1990). However, as only 10 specimens have been found in samples of this study, there is not enough data to designate a new species. It is most likely that *B. sp*. pollen grains are a variant of *B. rotatus*. Saccate living plants exhibit significant variations. On the other hand, *B. cf. rotatus*

Fig. 2.45 *Barakarites cf. rotates* showing proximal face with trilete scar

closely resembles *Cordaitina rotata.* Medvedeva (1960) described in detail from the Kungurian of the Perm region in Russia (Zavialova et al. 2004).

Barakarites rotatus (Bharadwaj and Tiwari 1964) (Fig. 2.46).
Occurrence: Lower to Upper Permian. Frequent in Artinskian to Upper Permian
Description: Monosaccate, radially symmetrical. Amb circular to rounded-triangular. Corpus outline faint, circular (Fig. 2.46a, b). Tetrad mark faintly visible; laesurae simple, varying in length from about half to 3/4 corpus radius, often of unequal length. Distal annular tenuitas sub-equatorial, varying in width from 1 to 10 μ. Endexine (nexine) often folded along rims of distal tenuitas. Nexine thinner than ektexine. Ektexine detached equatorially on proximal face and slightly sub-equatorially on distal face. Saccus has usually a smooth outline, although radial folds are present in some specimens.
Dimensions: Diameter 92–130 μ, corpus diameter 80–120 μ (20 specimens measured).
Remarks: *Barakarites rotatus* was recorded and described from Collie Basin by Backhouse (1991).
Occurrence: Upper Permian in Collie–Muja Basins; Bore B/224-227′, 2125′; Bore D/201-213′.
Remarks and discussion: *Barakarites* sp. Differs from *Barakarites* rotatus in possessing a thick nexine while *B. rotatus* is characterised by an extremely thin nexine. The thick nexine in *Barakarites* sp. gives the corpus a very dark appearance, while the thin nexine in *B. rotatus* makes the corpus appear hyaline. However, as only 10 specimens have been found in samples of this study, there is not enough data to designate a new species. It is possible that *B. sp.* pollen grains are a variant of

B. rotatus. Saccate living plants exhibit significant variations in the character of their sacs. On the other hand, *B. sp*. closely resembles *Cordaitina rotata* described in detail from the Kungurian of the Perm region in Russia (Zvialova et al. 2004).

Genus ***Parasaccites*** (Bharadwaj and Tiwari 1964).
Diagnosis: Monosaccate, more or less radially symmetrical with a varied tetrad mark. Nexine thinner or thicker than ektexine, tends to fold under saccus root. Ektexine detached equatorially on proximal face and sub-equatorially on distal face. Structure of ektexine intra-granulate to micro intra-reticulate.
Known distribution: Permian of Western Australia (Balme and Hennelly 1952, 1956), Permian of New South Wales (Balme and Hennelly 1956), Permian of India (Virkki 1945; Mehta 1944; Potonie and Sah 1960; Potonie and Lele 1961, Bharadwaj 1962; Bharadwaj and Tiwari 1964; Banerjee and D'Rozario 1990; Mishra and Jha 2017). Permo-Carboniferous glacials of Uruguay (Machiavello 1963), Permian of Karroo series, Tanganyika (Hart 1960, 1963), Permo-Carboniferous of Congo (Hoeg and Bose 1960).
Discussion: *Parasaccites* was established by Bharadwaj and Tiwari (1964) for monosaccate, more or less radially symmetrical pollen forms possessing a varied tetrad mark. Similar forms were described under *Nuskoisporites* (Potonie and Klaus and others). Bharadwaj and Tiwari (1964) carried out a detailed study of cf. *Nuskoisporites* forms and separated the Gondwana forms from the European *Nuskoisporites* mainly on the basis of the 'para-condition of saccus attachment'. On studying Klaus's (1963) detailed description accompanied by excellent illustrations and drawings, as well as observing specimens supplied by Dr. Klaus, I did not find any difference in the mode of ektexine detachment between the two genera. It is the same in both *Nuskoisporites* and *Parasaccites*, namely equatorially on proximal face and sub-equatorially on distal face. This is further confirmed from observing the microtome sections of *Parasaccites* pollen grains. Nevertheless, the present study yielded sufficient evidence in justifying the separation of the two genera. The main difference at first sight is the presence of a *limbus* sensu Potonie and Kremp (1955). In *Nuskoisporites*, this *limbus* is responsible for the umbrella-shaped forms so well demonstrated in the laterally compressed specimens of this pollen. However, no laterally compressed specimens were found in samples of the present study among hundreds of *Parasaccite*s specimens observed. Other distinct features are the thick and rigid nexine, and the coarse structure of the ektexine with well-developed columellae in *Nuskoisporites*, and their absence in *Parasaccites*.

Parasaccites gondwanensis (Balme and Hennelly 1956) (Fig. 2.47).

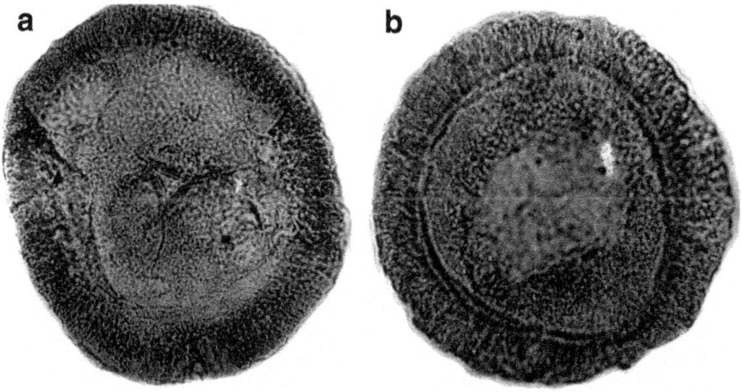

Fig. 2.46 **a** Proximal view, **b** distal view, partly missing

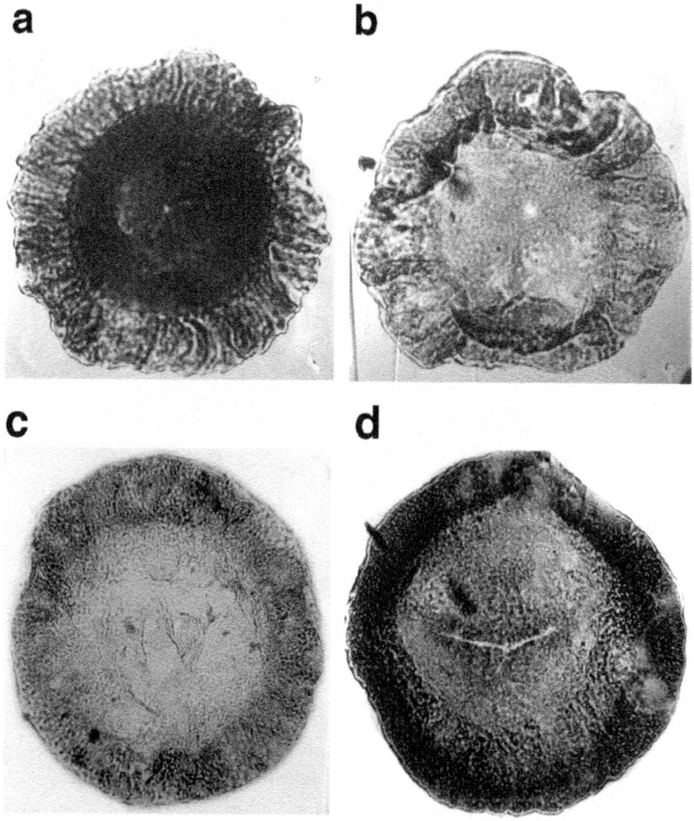

Fig. 2.47 **a** Corpus present, thick nexine, tetrad mark incomplete. **b** Corpus mostly missing, ektexine present, faint trilete mark detectable. **c** Corpus (nexine) missing. **d** Incomplete tetrad mark. Half corpus (ektexine) absent

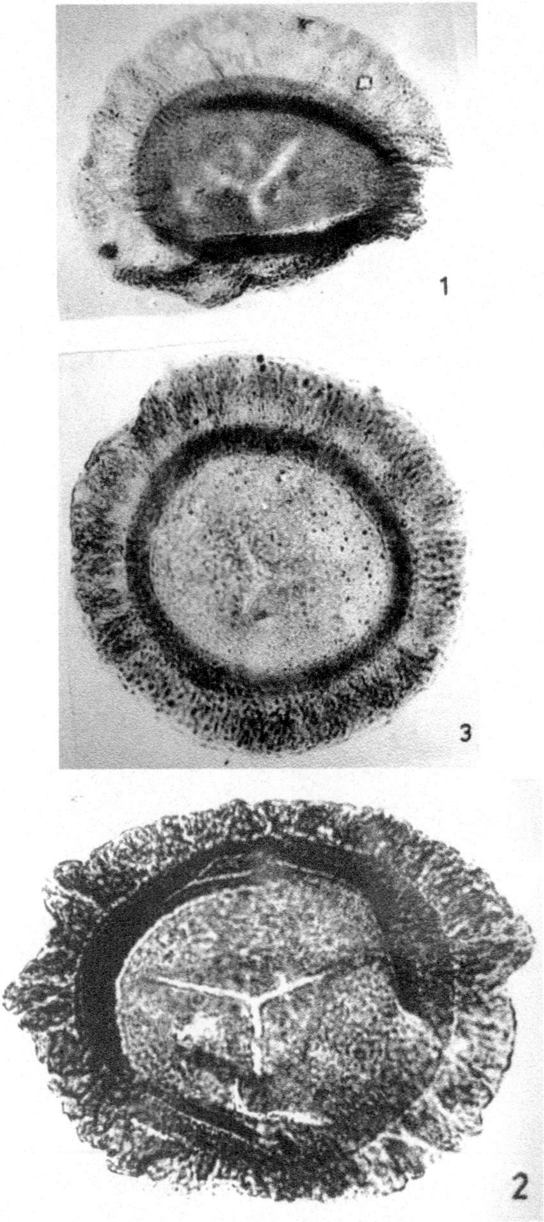

1–3 Show the typical circum-circular fold, evident in
microtome cross-section (see reconstructed specimen)
'complete' pollen, clear tetrad mark, nexine fold prominent

Occurrence: Lower Permian of Collie–Muja and Wilga Basins. In the Sakmarian of Sue No. 1 bore in Perth Basin it reaches 20%, and drops to 6% in the Lower

Artinskian. It is the most significant component of the Lower Permian, a major component, and often sole component of Sakmarian tillite pollen. It can be called a marker pollen.
Description: Monosaccate, radially symmetrical. Amb circular to sub-circular. Corpus circular. Tetrad scar clearly visible where corpus is preserved. Trilete, bilete or monolete. In trilete specimens scar usually not reaching equatorial margin. Ektexine detached equatorially on proximal face, and sub-equatorially on distal face. Endexine/nexine thicker than ektexine. The latter intensely radially folded, the result of a flattened highly inflated saccus. Thin sections of the pollen in the present study confirmed the folding over of the corpus nexine. Structure of ektexine intra-granulate.
Dimension: Total diameter 40–140 μ. Corpus diameter 36–120 μ (100 specimens measured).
Remarks and discussion: On examining hundreds of specimens of *Parasaccites* (*gondwanensis*) from the same stratigraphic horizon, gradations were found from specimens possessing a complete annular sub-equatorial fold to a partial fold. Also, gradations from a thick rigid corpus exine to a thin corroded exine, to absence of a corpus altogether. Gradations from inaperturate specimens through monolete and bilete to perfect trilete have been observed in the present study. Nygreen and Bourne (1967) stressed the importance of recognising transitions among the forms rather than try and separate as many as possible on the basis of minor differences.

Specimens with *Parasaccites* characteristics have been assigned to other genera, such as *Plicatipollenites* in Permian of Collie (Backhouse) and Permian of India (Banerjee and D'Rozario). The stratigraphic and climatic spread of '*Plicatipollenites*' and '*Parasaccites*' in the Permian of Chuparbita Basin, India (Banerjee and D'Rozario) is interestingly the same as in the Collie Basin. Therefore, it would be useful to combine the two forms under one genus; either *Parasaccites* or *Plicatipollenites*.

Reconstructed specimen of *Parasaccites gondwanensis* (Fig. 2.47/4) on p.68.

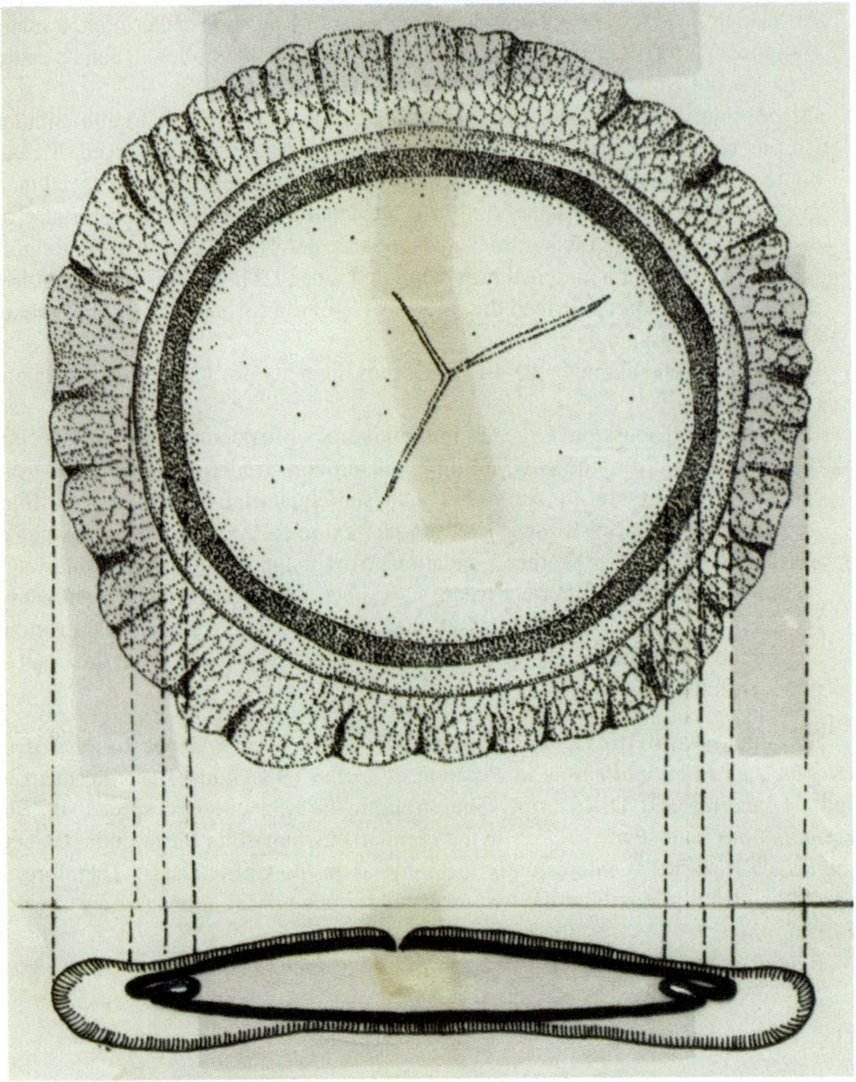

Genus ***Potoniesporites*** (Bharadwaj 1964a, b).

Diagnosis: Monosaccate, bilaterally symmetrical. Amb oval to sub-circular. Monolete; proximal scar not always clearly visible. Corpus outline circular, sub-circular or rhomboidal. Corpus ektexine smooth or structured, detached equatorially on proximal face and intra-equatorially on distal face.

Known distribution: Upper Carboniferous to Permian of Northern and Southern hemispheres.

Discussion: *Potoniesporites* was first described from the Carboniferous of the Saar. Bharadwaj described the genus first in 1954, and emended it in 1964 giving a more detailed description and emphasising the 'cappula' or 'distal zone' in

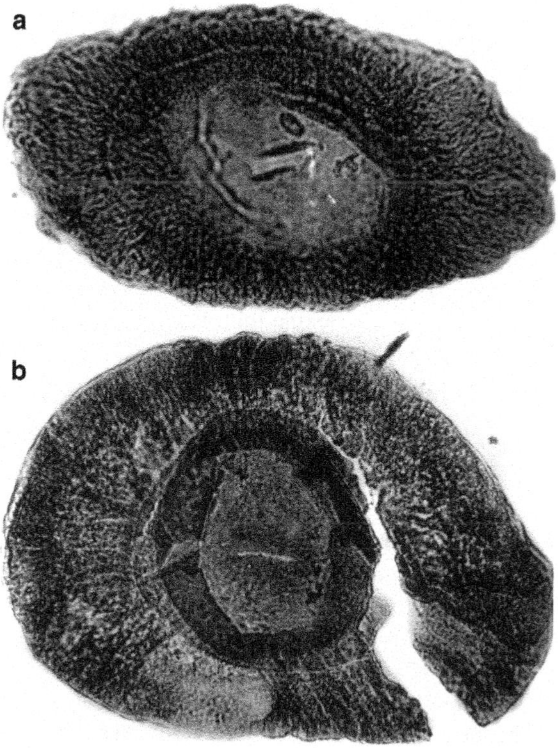

Fig. 2.48 **a** Distal face; outline of corpus mostly visible. Saccus roots forming a semi-circular 'sulcus'. **b** Proximal face with a partial 'tetrad' scar

Potoniesporites. The cappula was regarded by Bharadwaj as a most distinct feature, usually bordered by folds. The emended form could include other monosaccate genera, described previously, such as *Vestigisporites Balmei* reported by Hart (1960) from the Lower Permian coal measures of Tanganyika. *Crucisaccites* (Lele and Maithy 1964) closely resembles *Potoniesporites* and is probably congeneric with it. *Sahnites* (Pant 1955) is regarded in the present study as congeneric with *Potoniesporites*. Backhouse (1991) kept the two genera separate.

Potoniesporites neglectus (Potonie and Lele 1951) (Fig. 2.48).
Occurrence: Sakmarian to Upper Permian. More common in Sakmarian of Collie Basin.
Description: Monosaccate, bilaterally symmetrical. Monolete. Scar varies in length from a few microns to almost entire width of corpus, occasionally followed by folds. Corpus oval to sub-circular in outline. Nexine smooth and thin. Saccus ektexine detached equatorially on proximal face and sub-equatorially on distal face, coarsely structured; intra-reticulate.
Dimensions: 74–180 μ (20 specimens measured).
Remarks: *P. bilateralis* closely resembles *P. neglectus* (Potonie and Lele 1951), the sole difference pointed out by Singh (1964) being the 'trapezoidal' corpus and finer structure. However, both features are not significant enough to separate

the species. The 'trapezoidal' look is likely to be a result of the angle of preservation of the pollen grain. Backhouse (1991) recorded *Potoniesporites* from Collie Basin under *P. balmei* (Hart) and *P. novicus* (Bharadwaj). *P. balmei* appears close in shape and form to *Potoniesporites neglectus* specimen in Fig. 2.48a; while *P. novicus* appears very close in shape and form to *P. neglectus* specimen in Fig. 2.48b.

Sub Turma: Disaccites (Cookson 1947; Potonie and Klaus 1954).
Genus ***Sulcatisporites*** (Leschik 1955; Bharadwaj 1962).
Diagnosis: Disaccate, with occasional laterally connecting ektexinal strip. Amb circular to oval. Nexine thin. Rendering the corpus outline not easily detected. Sacci distally offset. Cappula narrow.
Known distribution: Artinskian to Mesozoic of the Northern and Southern hemispheres.
Discussion: Bharadwaj on emending *Sulcatisporites* gave a detailed description of the genus and noted the absence of a sulcus in the true sense. The narrow distal gaps between the sacci correspond to a cappula. The sacci overlap most of the distal face of the pollen grain, leaving a very narrow cappula. Hart (1965a, b) treated *Sulcatisporites* as synonym of *Vesicaspora. Vesicaspora* possesses a significantly thicker nexine than *Sulcatisporites.*
Several other monosaccate pollen grains such as *Alisporites, Pavisporites* and *Rimasporites* are regarded as synonyms of *Sulcatisporites* in the present study. *Protoconiferus* (Bolchovitina 1956) is described in detail and proves to be congeneric with *Sulcatisporites.* Bolchovitina uses the term 'furrow' for cappula. However, *Sulcatisporites* is senior to *Protoconiferus,* although its description is not as detailed; nevertheless, it has priority over *Protoconiferus.*

Sulcatisporites potoniei (Lakhanpal, Sah and Dube 1958) (Fig. 2.49).
Occurrence: Artinskian to Upper Permian. More abundant in Artinskian of Collie Basin.
Description: Disaccate. Amb circular, sub-circular or oval. Diploxylonoid. Sacci occasionally connected equatorially by a narrow exinal strip. Nexine very thin responsible for the vague outline of corpus. Corpus varied in size; ratio of corpus to sacci varies. Sacci overlap corpus distally almost entirely, leaving a narrow cappula, usually not exceeding 2 μ in width.
Structure intra-punctate on attached ektexine and intra-reticulate on sacci.
Dimensions: Total width 75–115 μ. Corpus width: 30–60 μ (20 specimens measured).
Remarks and comparison: *Sulcatisporites reticulatus* (Madler 1964) from the Triassic of Germany closely resembles *S. potoniei. S. kraeiseli* also from the Triassic of Germany is slightly smaller in size than *S. potoniei* and has a wider cappula.

Sulcatisporites ovatus (Balme and Hennelly 1962) (Fig. 2.50).
Occurrence: Common in Upper Permian samples of Collie–Muja Basin and Alexandra Bridge Bore.

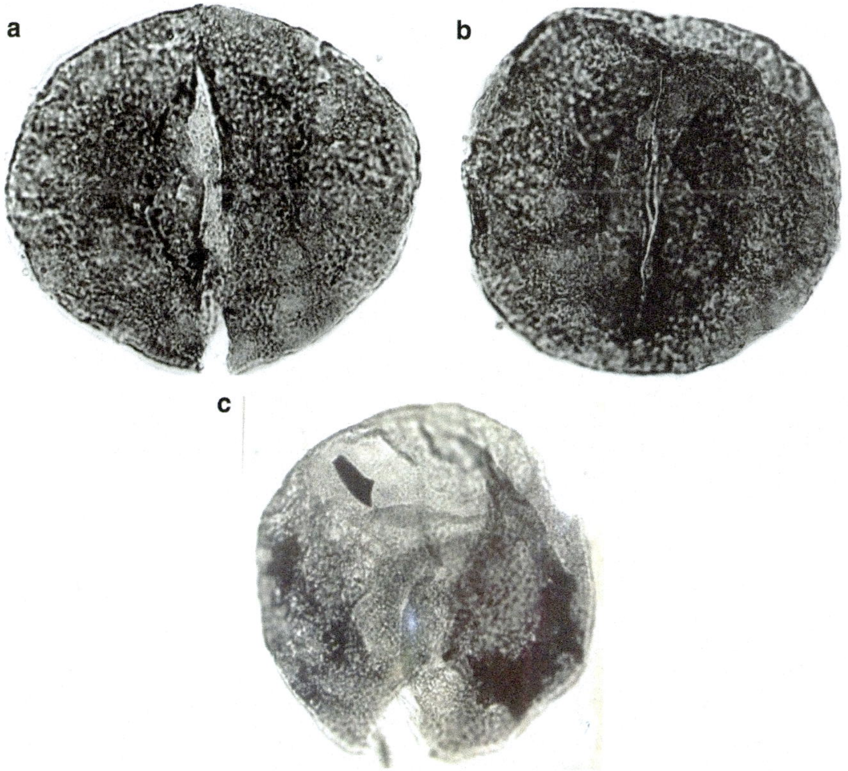

Fig. 2.49 **a**, **b** Faintly visible corpus outline. Sulcus usually pronounced **a** and **b** distal face showing sulcus, while corpus only faintly visible. **c** Side view showing proximal face very finely structured in contrast to coarse structure on distal face/sacci

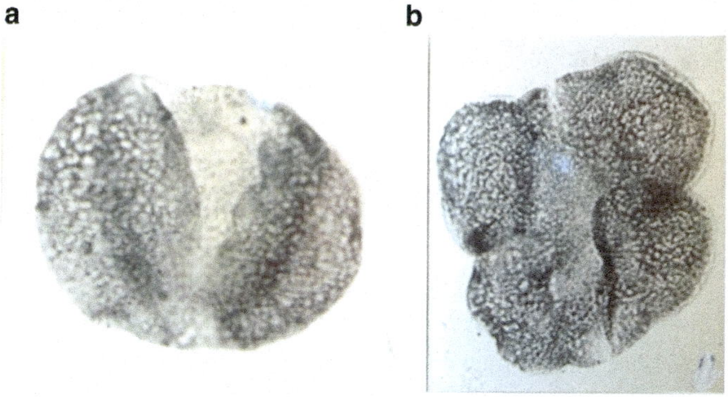

Fig. 2.50 **a**, **b** tetrad

Description; Haploxylonoid to slightly diploxylonoid. Small in size. Occasional tetrads observed (Fig. 2.50). Sacci slightly smaller than corpus, sub-circular to oval. Outline of corpus sub-circular to oval, not sharply outlined due to thin nexine. Ektexine detached equatorially on proximal face and sub-equatorially on distal face, leaving a narrow cappula, a distinct feature of the genus. Structure of attached ektexine finely intra-punctate. Sacci finely intra-granulate to intra-microreticulate.
Dimensions: 30–50 μ (25 specimens measured).
Remarks: *Sulcatisporites ovatus* is distinguished from *S. potoniei* by its distinctly small size, and much finer structure.

Genus ***Platysaccus*** (Naumova) (Potonie and Klaus 1954).
Diagnosis: Disaccate, diploxylonoid. Sacci larger than corpus. Corpus outline sub-circular. Overall a large pollen.
Known distribution: Palaeozoic of Northern and Southern hemispheres.
Discussion: *Platysaccus* was first described by Naumova (1937) from Russian Carboniferous coals. Potonie and Klaus (1954) validated the genus in naming a type species. *Platysaccus* is distinguished by its large sacci, making it look like a butterfly, further emphasised by the shape of the corpus being usually elongated. The large sacci, clearly outlined corpus and overall size distinguishes *Platysaccus* from *Alisporites* and *Illinites.*

Platysaccus major (Hoeg and Bose 1960) (Fig. 2.51).
Occurrence: Artinskian to lower part of Upper Permian of Collie–Muja Basin.
Description: Disaccate, diploxylonoid. Large pollen (average 120 μ total width). Sacci larger than corpus. The large sacci always folded on distal face forming a sulcus-like image. Corpus clearly outlined, sub-circular to elongated. Ektexine detached equatorially on proximal face and sub-equatorially on distal face. Cappula narrow. Endexine varies in thickness; as thick or thicker than ektexine, folds under distal roots. Structure of ektexine intra-punctate on attached parts and intra-micro-reticulate on sacci.
Dimensions: 90–130 μ width of pollen (15 specimens measured). Remarks: *P. queenslandii* (de Jersey) has a wide cappula and a 'sulcus'.

Platysaccus leschikii (Hart 1960) (Fig. 2.52).
Occurrence: Lower Permian of Collie–Muja–Wilga basins. Rare in Upper Permian.
Description: Disaccate, Diploxylonoid. Outline of corpus circular. Sacci very large in relation to corpus. Small corpus appears as if being 'wrapped up' by sacci. Corpus: sacci ratio is 1:4.
Endexine/nexine thicker than ektexine. Ektexine detached equatorially on proximal face and intra-equatorially on distal face. Large wrapping sacci obscure cappula. Structure of corpus nexine appears to be verrucate. However, not clearly discernible due to much folded sacci cover. Saccus ektexine micro-reticulate.
Dimensions: Total width: 98–122 μ; corpus 30–40 μ across (15 specimens measured).

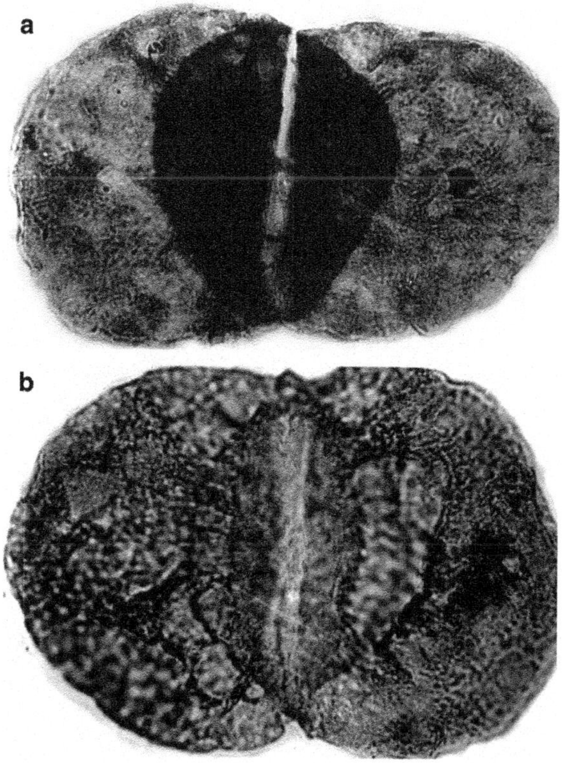

Fig. 2.51 **a** Distal view. Sacci folded along roots, **b** corroded specimen with nexine missing showing clear view of sacci roots

Remarks: *Platysaccus leschikii* is distinguished by the small corpus in relation to sacci. Sacci are wrapping corpus and giving it an almost monosaccate appearance. *Florinites eremus* described from Collie Basin by Backhouse (1991) resembles *P. leschikii*, but seems to have a more pronounced monosaccate appearance. The resemblance is in the character of the corpus, being small and rigid within a large originally highly inflated sac/sacci.

Genus ***Illinites*** (Kosanke 1950).
Diagnosis: Disaccate pollen. Diploxylonoid to haploxylonoid. Cappa possesses a tetrad scar. Endexine as thick or thicker than ektexine.
Known distribution: Upper Carboniferous of North America (Kosanke 1950), Upper Permian of Germany (Potonie and Klaus 1954; Leschik 1956; Grebe 1957; Klaus 1963); Permian of Tanganyika Hart 1960).
Discussion: *Illinites* was originally described from Upper Carboniferous (Kosanke 1950). Kosanke restricted the genus to forms possessing a trilete mark on their proximal face. However, a trilete mark is not always complete. The present study shows that the rays may be uneven, or one ray missing, making it bilete, or even monolete. Therefore, a trilete mark should not be used as a distinguishing feature on a generic level. Hence, *Limitisporites* and *Jugasporites* both proposed

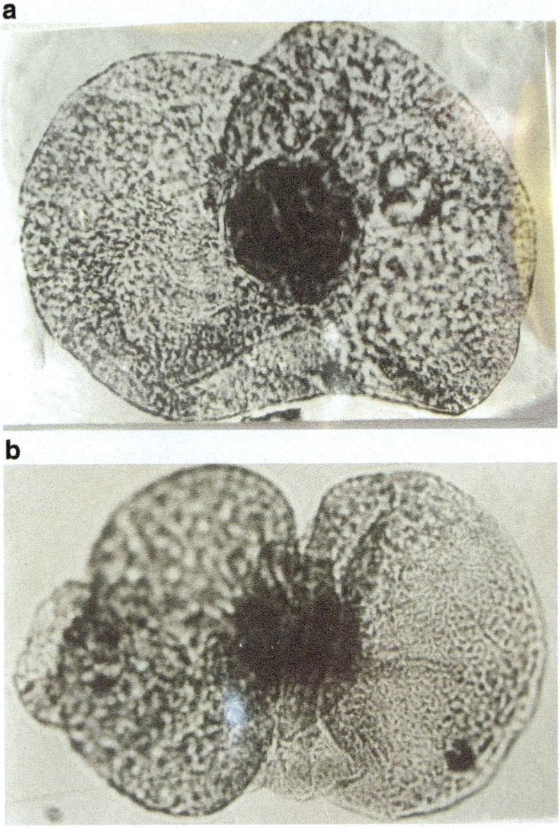

Fig. 2.52 **a** Proximal face; small corpus in relation to sacci, typical feature. **b** Distal face

by Leschik (1956) for monolete and dilete disaccates, respectively, are regarded here as congeneric with *Illinites*.

Schaarschmidt (1963) compared *Limitisporites* and *Illinites* with spores of *Ulmania frumentaria* (Potonie and Schweitzer 1960), emphasising the variations of the proximal scar, from monolete to dilete and trilete. However, Schaarschmidt combined *Limitisporites* and *Jugasporites* under one genus, maintaining *Illinites* as a separate genus. Schaarschmidt's reason for maintaining two genera for similar pollen grains were purely stratigraphical, namely that one was described from the Upper Carboniferous of North America, while the other two were described from the Permian of Europe. The present study maintains one genus.

Illinites tectus (Leschik 1956) (Fig. 2.53).

Occurrence: Artinskian of Collie and Wilga Basins.

Description: Disaccate. Haploxylonoid to slightly diploxylonoid. Sacci vary in size, smaller than corpus or same size, occasionally slightly larger. Corpus very distinct in its regular shape, size and exine thickness. Monolete, scar not always

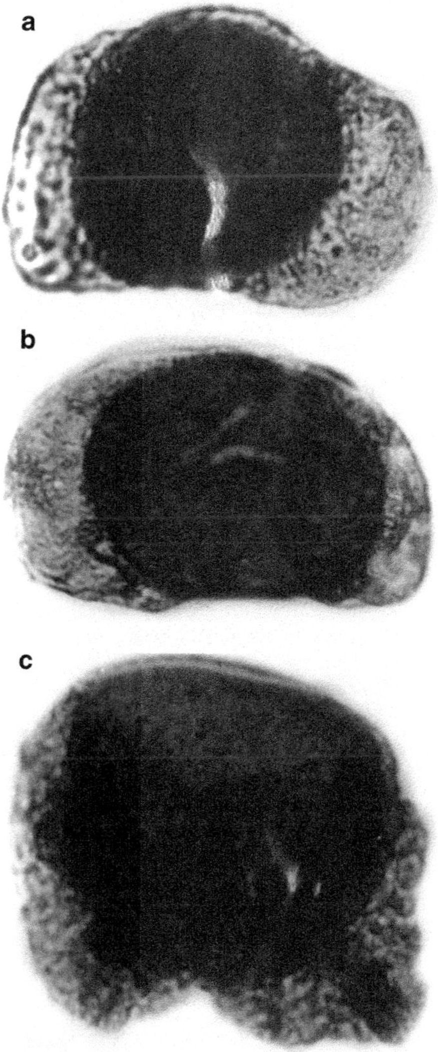

Fig. 2.53 **a** Proximal face; thick nexine of corpus, sacci about of same size as corpus, **b** distal face, **c** side view showing sacci position

clear, often faintly outlined. Endexine significantly thicker than ektexine, responsible for the dark corpus. Ektexine detached equatorially to slightly sub-equatorially on proximal face and sub-equatorially on distal face. Cappula 1/3 corpus width. Structure of sacci fine, intra-granulate to intra-microreticulate.

Attached ektexine non-structured. Sacci occasionally joined laterally by an ektexinal strip.

Dimensions: Total width: 51–84 μ; corpus: 24–54 μ (25 specimens measured).

Remarks and comparison: *Illinites tectus* is distinguished by its thick and rigid nexine and finely structured thin sacci. *I. tectus* has no nexine folds under distal roots, a phenomenon common in other species of *Illinites*.

Genus ***Alisporites*** (Daugherty) (Nilsson 1958).
Diagnosis: Disaccate, haploxylonoid to slightly diploxylonoid. Sacci distally inclined, same size or slightly larger than corpus. Nexine folded under distal roots. Cappula tenuinexinous.
Known distribution: Widespread in Permian in Northern and Southern hemispheres.
Discussion: The form genus *Alisporites* was erected and used by Daugherty (detailed in Potonie and Kremp 1956) for large disaccate pollen with a distal sulcus, 'a single fusiform furrow' as described by Daugherty. Potonie and Kremp (1956) designated a type species, namely *Alisporites opii*. Potonie and Kremp followed Daugherty in interpreting the torn cappula as a sulcus. Nilsson (1958) was not in the opinion that Alisporites possesses a sulcus in the true sense of a sharply delimited furrow. Nilsson observed in similar forms a distal tenuitas and on occasion exine was missing altogether due to bad preservation.
Sulcatisporites is distinct from *Alisporites* by its vaguely distinguishable corpus and narrow cappula.
Platysaccus has larger sacci than corpus.

Alisporites milvinus (Balme and Hennelly 1955) (Fig. 2.54).
Occurrence: Upper Permian of Collie–Muja Basin.
Description: Disaccate. Diploxylonoid. Sacci as large as or smaller than corpus. Ektexine detached equatorially on proximal face and sub-equatorially on distal face. Cappula about 1/3 corpus width, tenuinexinous. Exine thick, nexine folds under distal roots, forming a sharp outline for cappula. Exine smooth on attached parts and intra-microreticulate on sacci.
Exine tends to split along the central, presumably thinnest part of cappula.
Dimensions: Total width: 60–75 μ. Corpus width: 30–54 μ (10 specimens measured).
Remarks: *Alisporites* closely resembling the Collie Basin species was recorded from the Permian of Antarctica (Faebee et al. 1990).

Sub-turma: *Striatites* (Pant 1954).
Pollen grains belonging to the Striatites, having characteristic taeniate/striate corpus, have been isolated from sporangia of the Glossopterids (Lindstrom et al. 1997). The glossopteris flora is the dominant group of plants in the Permian sediments, and a major contributor to the coal deposits of Gondwana. Glossopterids are a cold climate vegetation and shed their leaves in winter. McLaughlin (1992) stated that "different parts of the plants were often fossilised separately. As a result many different names have been applied to various fossilised organs of this group." The same applies to the pollen of the Glossopterids. In this book I make an attempt in combining names of pollen to what appears to be the same plant by comparing the fossil pollen/spore to genera of similar living plants. This is

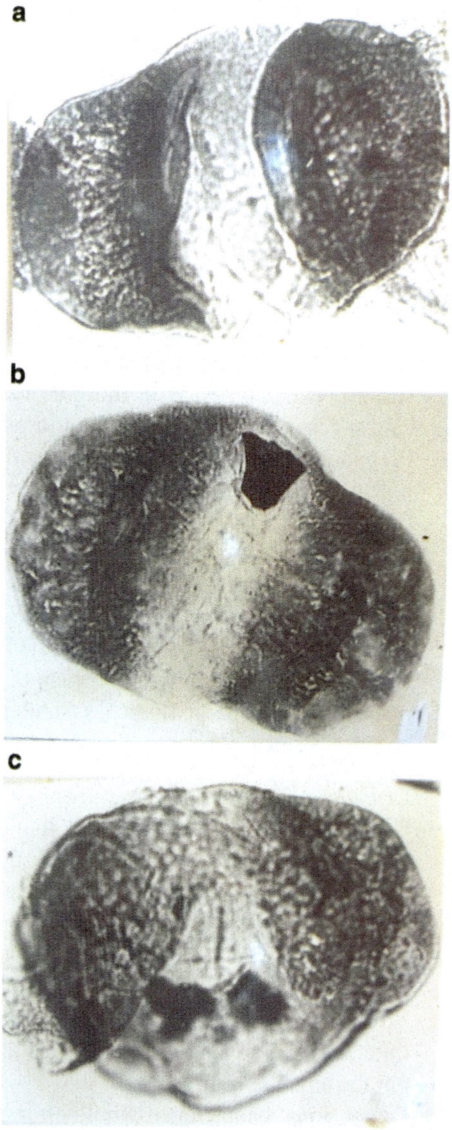

Fig. 2.54 **a** Distal view; tear in area between sacci misinterpreted as sulcus. **b** Thin cappula and sacci roots. **c** Side view showing position of sacci in relation to corpus

possible with all saccate fossil pollen. However, the Striatites have no equivalent in living plants since the Glossopterids are an extinct group, and therefore no direct comparison with living genera is possible. Nevertheless, a general view of diversity can be applied, and an attempt at reducing the number of species is made.

Genus ***Crustaesporites*** (Leschik 1956; Jansonius 1962).

Diagnosis: Monosaccate. Amb circular to lobed. Proximal face taeniate. Ektexine detached equatorially on proximal face and sub-equatorially on distal face. Outline of corpus circular. Nexine folded under distal root.
Known distribution: Upper Permian of Germany (Leschik 1956; Grebe 1957), Upper Permian of England (Clark 1965). Upper Permian of India (Bharadwaj 1962), Lower Triassic of Canada (Jansonius 1962), Permian of Midcontinent USA (Jizba 1962), Lower Triassic of Western Australia (Balme 1963).
Discussion: *Crustaesporites* was originally described by Leschik as a trisaccate form. Jansonius believed the pollen to be monosaccate-lobed. It was noticed in the present study that variations and gradations occur, from monosaccate forms to bi-lobed, tri-lobed and poly-lobed.
Hart suggested that taeniate-monosaccate forms are monstrosities of disaccate striatiti. It is suggested in the present study that they are variants of a bisaccate species, as is not uncommon in pollen of living trees, descendant from the Permian flora, that is, *Podocarpus papuanus* has bilobed to poly-lobed saccate pollen. Similarly, pollen of *Podocarpus gracilior* can be bisaccate to trisaccate (Fig. 4). Cousminer (1965) also treats *Crustaesporites* as a polysaccate form, by comparing the fossil pollen to present-day existing genera such as *Podocarpus*, *Microcarpus* and *Microstrobus* which all show transitions from polysaccate to lobed to monosaccate.
Crustaesporites is rare in Collie–Muja Basin. Monosaccate or lobed striatiti seem to be common in the Upper Permian of India (Bharadwaj 1962) which justifies a place for them in palynological systematics. Hart (1965a, b) treats *Crustaesporites* as a polysaccate form.

Crustaesporites ovatus (Bharadwaj 1962) (Fig. 2.55).
Occurrence: Upper Permian of Collie–Muja Basin.
Description: Lobed monosaccate or trisaccate. Amb circular to lobed. Corpus outline circular to sub-circular. Ektexine detached equatorially on proximal face and sub-equatorially on distal face. Proximal face of corpus taeniate. Taeniae 2–4 μ wide, often wedge out or are interrupted by vertical striae. Structure of attached ektexine on proximal and distal faces intra-punctate to intra-granulate. Detached ektexine intra-reticulate. Reticulum consists of brochi up to 8 μ across. Nexine thinner than ektexine, folds under distal root. Exine on distal face of corpus thins out towards the centre.
Dimensions: Diameter of whole grain 65–135 μ. Corpus diameter range 40–60 μ (10 specimens measured).

Genus ***Striatopodocarpites*** (Sedova 1956; Hart 1964).
Diagnosis: Disaccate striatiti. Diploxylonoid. Sacci distinctly larger than corpus. Cappa taeniate; divided into more than four taeniae. Taeniae continuous or broken, parallel or converging. Corpus outline circular or oval.
Known distribution: Widespread in the Permian of the Northern and Southern hemispheres.

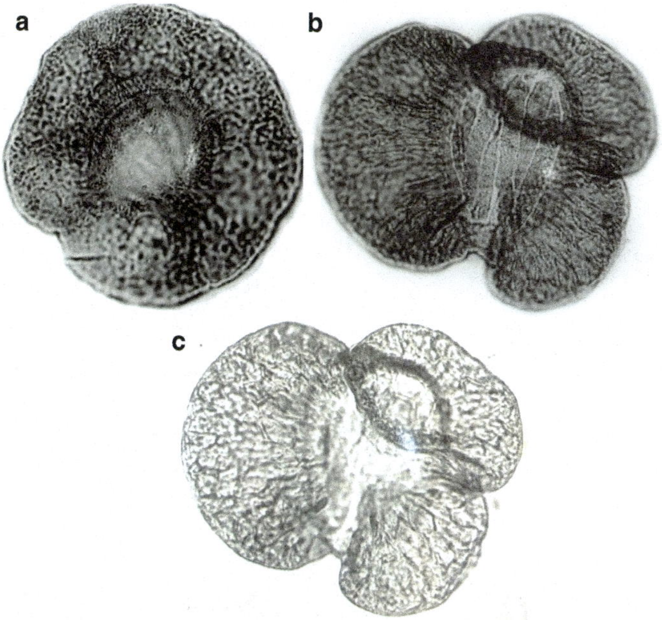

Fig. 2.55 **a** Distal view of a monosaccate specimen, **b** proximal taeniate view of lobed specimen, **c** distal view

Discussions: *Striatopodocarpites* was first described and illustrated by Sedova (1956).
S. tojmensis was designated as type species. Hart (1964) emended the genus to accommodate proximally striated diploxylonoid forms with sacci larger than corpus.

Striatopodocarpites gondwanensis (Lakhanpal, Sah and Dube 1958) (Fig. 2.56).
Occurrence: Upper Permian of Collie–Muja Basin and Alexandra Bridge bore.
Description: Strongly diploxylonoid. Sacci about twice as large as corpus. Ektexine detached equatorially on proximal face and sub-equatorially on distal face. Cappula no more than half the corpus width. Cappa striated; Taeniae number about 10, diverging occasionally as well as breaking up by vertical grooves or striae. A proximal equatorial ridge of loose ektexine is present in some specimens. Nexine thicker than ektexine, up to 3 μ thick in some specimens, occasionally folded under distal roots. Structure of ektexine coarse, intra-punctate on attached parts and dense intra-reticulate on sacci.
Dimensions: Total width: 70–140 μ; corpus width: 35–50 μ (20 specimens measured).
Remarks and comparison: *S. gondwanensis* was first described by Lakhanpal et al. (1959) from the Permian of India. This species of *Striatopodocarpites* is characterised by its discontinuous proximal taeniae which are broken up by vertical

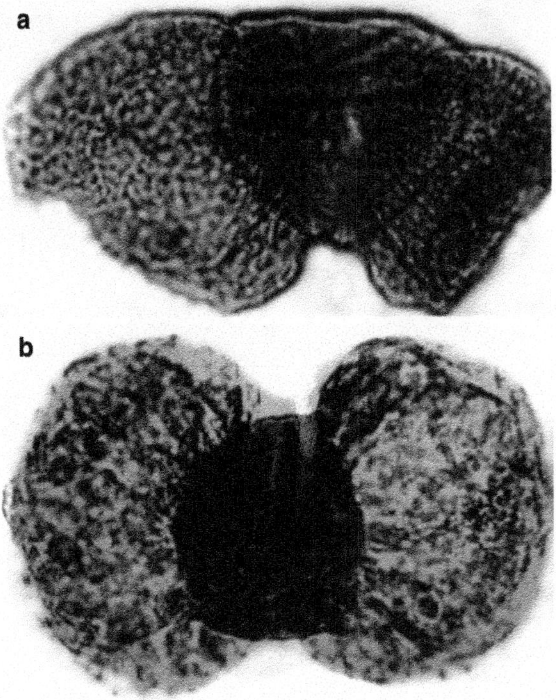

Fig. 2.56 **a** Semi-distal view, sacci roots. **b** Proximal view, coarse taeniae

striae. Other specific characteristics are large sacci, narrow cappula and thick nexine. Structural elements coarse. A proximal equatorial ridge occasionally observed is not considered a specific feature since it is found in other disaccate forms as well as in existing plants' pollen.

Striatopodocarpites cancellatus (Balme and Hennelly) (Hart 1964) (Fig. 2.57).
Occurrence: Upper Permian of Collie–Muja Basin, Alexandra Bridge bore.
Description: Disaccate striatiti. Diploxylonoid. Sacci larger than corpus. Corpus outline sub-circular. Cappa possesses 5–6 taeniae which often wedge out. Taeniae vary in width from 2 to 6 μ, and are intra-punctate to intra-granulate in structure. Ektexine detached equatorially on proximal face and sub-equatorially on distal face. Occasional 'equatorial ridge' present. 'Ridge' consists of loose ektexine. Cappula 1/3 to 1/4 corpus width. Nexine thicker than ektexine. Sacci structure fine, intra-reticulate.
Dimensions: Total width: 35–76 μ; corpus 18–34 μ across (14 specimens measured).
Remarks and comparisons: *S. cancellatus* was originally described from the Permian of New South Wales (Balme and Hennelly 1955) and is distinguished by its thick corpus, thinner sacci and fine structure. *S. gondwanensis* is significantly larger and its proximal taeniae are broken up by vertical partitions. *Striatopodocarpites fusus* has heavier sacci and is larger in size.

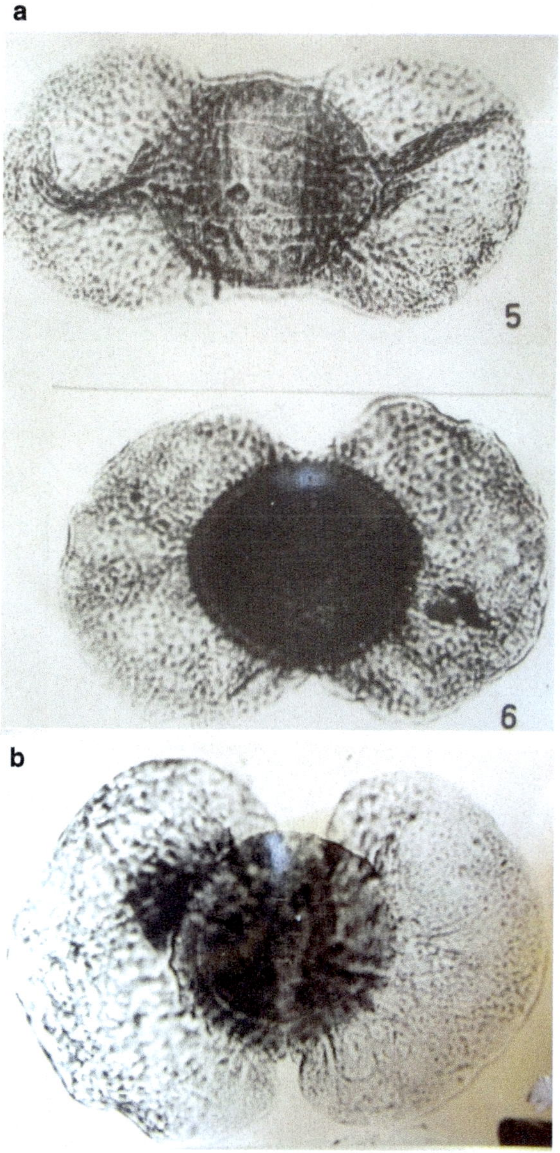

Fig. 2.57 **a** Taeniae of corpus. **b** Thick nexine of corpus. Distal view: 'sulcus', sacci roots

Genus ***Protohaploxypinus*** (Samoilovich 1953; Hart 1964).
Diagnosis: Disaccate striatiti. Haploxylonoid to slightly diploxylonoid. Corpus outline circular. Cappa striated; divided into four or more taeniae. Sacci of same size as corpus or slightly smaller.
Known distribution: Lower Permian to Triassic of Northern and Southern hemispheres. Reported from Permian coal of Brazil (Cazzulo-Klepzig et al.2009), from Permian of Antarctica (Farabee et al. 1990), Permian of Collie Basin (Backhouse 1991).

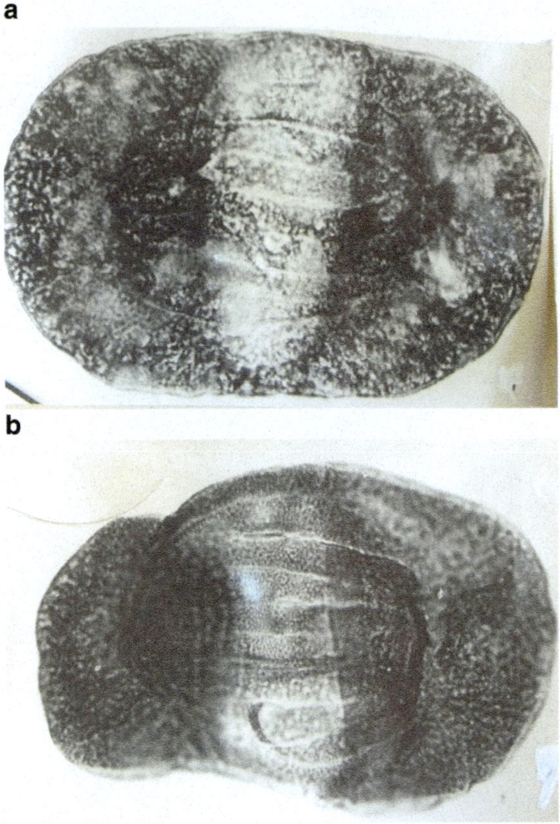

Fig. 2.58 **a** and **b** Showing sacci roots and proximal taeniae

Discussion: *Protohaploxypinus* is distinguished from *Striatopodocarpites* by the smaller sacci. The latter are of smaller or same size as corpus, whereas the sacci in *Protohaploxypinus* are significantly larger than corpus; which translates into balloon-like, blown out in the pollen before sedimentation and compaction. *Protohaploxypinus* has priority over *Faunipollenites* (Bharadwaj 1962), otherwise could be the same genus. *Taeniaesporites* (Leschik) (Jansonius 1962) is distinguished by its restricted number of taeniae which are not less than three and do not exceed five.

Protohaploxypinus amplus (Balme and Hennelly 1955; Hart 1964) (Fig. 2.58).
Occurrence: Frequent in Upper Permian samples of Alexandra Bridge bore and Upper Permian of Collie–Muja Basin. Less common in Artinskian and rare in Sakmarian.
Description: Disaccate striatiti. Haploxylonoid to slightly diploxylonoid. Corpus outline circular. Sacci are as large as corpus, or slightly smaller. Ratio of saccus to corpus is about 2. Cappa striated; number of taeniae 7–9, parallel or bifurcating. Nexine as thick or thicker than ektexine and often folds under distal roots. Ektexine detached equatorially on proximal face and sub-equatorially on distal face. Cappula tenuinexinous and about half corpus width.

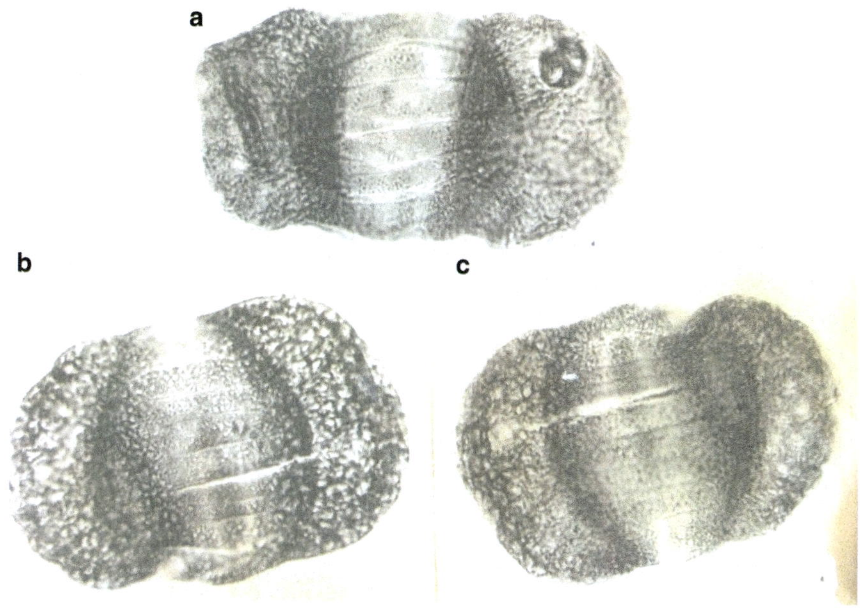

Fig. 2.59 **a** Taeniae and saccus roots. **b** and **c** Fine reticulate structure on endexine over corpus contrasts with coarse intra-reticulum on sacci

Structure of attached ektexine intra-punctate to intra-granulate. Sacci structure intra-microreticulate to intra-reticulate.
Dimensions: Total width: 70–116 μ; corpus: 40–54 μ across (100 specimens measured).
Remarks and comparison: *Protohaploxypinus amplus* was originally described from the Permian of New South Wales by Balme and Hennelly (1955). It is distinguished from *P. limpidus* as pointed out by Balme and Hennelly by its thicker ektexine and the larger size of *P. amplus*.
By measuring 100 specimens of each species in the present study, it is noted that the two forms grade into one another, mainly in corpus size, but also in total width. However, the two forms are separated here by drawing a size limit at about 65 μ total width, and 30 μ of corpus height/diameter. As can be seen in the graphs, there is an overlap between the two forms. Both *P. amplus* and *P. limpidus* share the same stratigraphic distribution. Therefore, in principle the amalgamation of the two forms into one species can be justified and is recommended.

Protohaploxypinus limpidus (Balme and Hennelly 1955) (Fig. 2.59)
Occurrence: Abundant in Upper Permian; Alexandra Bridge bore, Collie–Muja Basin. Less frequent in Lower Permian.
Description: Disaccate striatiti. Haploxylonoid; corpus outline sub-circular. Sacci slightly smaller than corpus. Cappa striated; number of taeniae 7–9. Ektexine equatorially detached on proximal face and intra-equatorially on distal face. Cappula varies in width from ¼ to ½ corpus width. Nexine thinner than ektexine, does

not fold under distal roots. Structure of attached ektexine is intra-punctate, and of sacci intra-microreticulate to intra-reticulate.

Dimensions: Total width: 40–70 μ; corpus height 27–48 μ (100 specimens measured).

Remarks and comparison: *P. limpidus* is distinguished from *P. amplus* by its size, being generally smaller. Comments to *P. amplus* apply, namely that *P. amplus* and *P. limpidus* ought to be amalgamated into one species.

Fig. 2.59 Corpus: sacci ratios

(1) Corpus height versus no. of specimen;

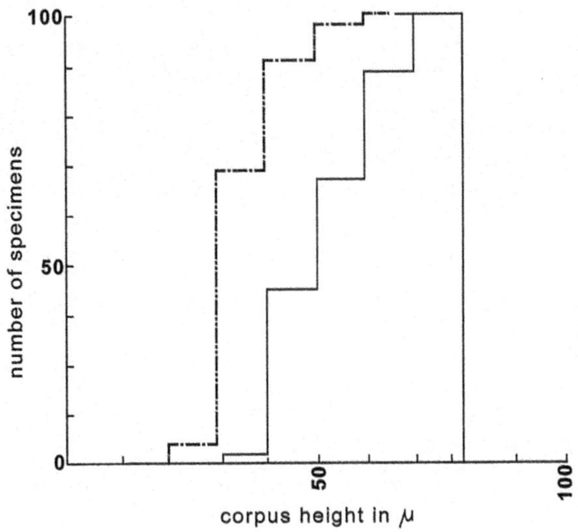

(2) Total width versus number of specimen;

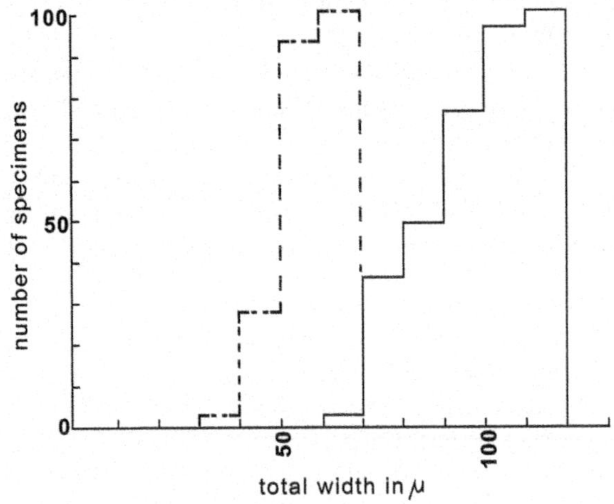

(3) Combined overlap of no of specimen and corpus height in *P. amplus* and *P. limpidus.*

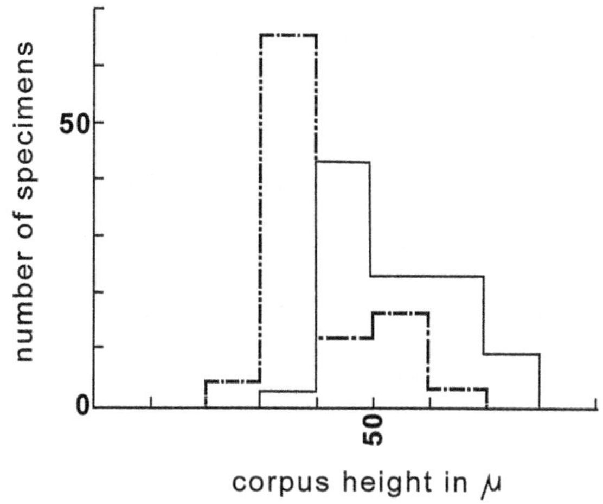

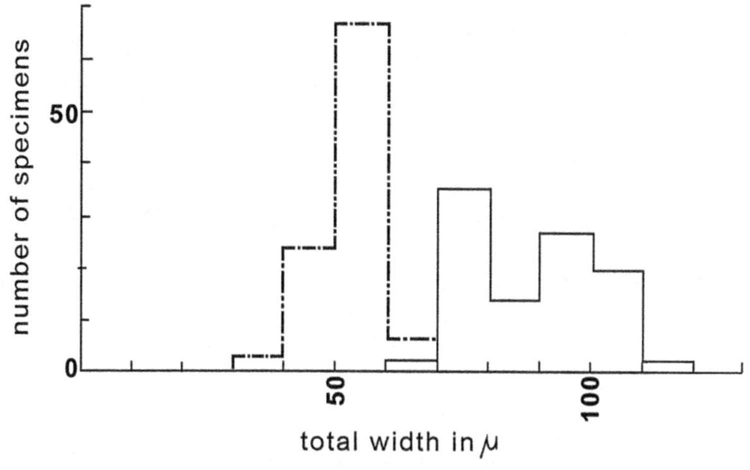

Protohaploxipinus diagonalis (Balme 1963) (Fig. 2.60).
Occurrence: Upper Permian of Collie–Muja Basin. Alexandra Bridge bore.
Description: Disaccate striatiti. Small, about 45 μ across (total width). Haploxylonoid. Sacci smaller than corpus. Corpus outline vaguely discernible. Circular to sub-circular cappa dissected by two to three oblique striae dividing taeniae. Taeniae oblique, not parallel, numbering three to four. Cappula narrow, 2–4 μ wide. Nexine very thin, often fold under distal roots. Ektexine thicker than nexine, finely structured; intra-punctate on attached parts and intra-micro-reticulate on sacci.
Dimensions: Total width: 35–55 μ (20 specimens measured).
Remarks and comparison: *Protohaploxypinus diagonalis* is distinguished from the other species of the genus by its diagonal-wide striae, small size and thin nexine.

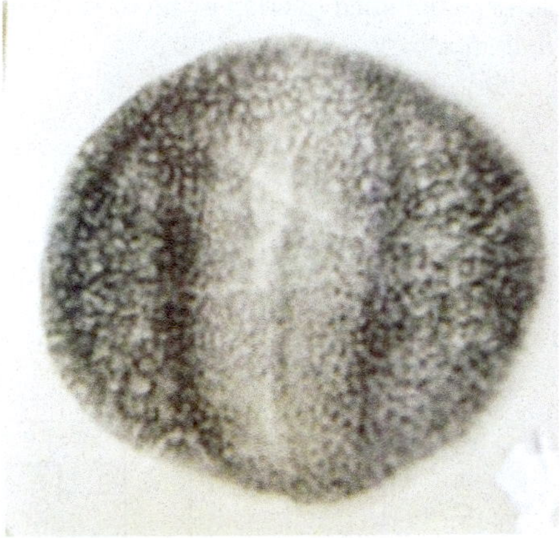

Fig. 2.60 *Protohaploxypinus diagonalis*

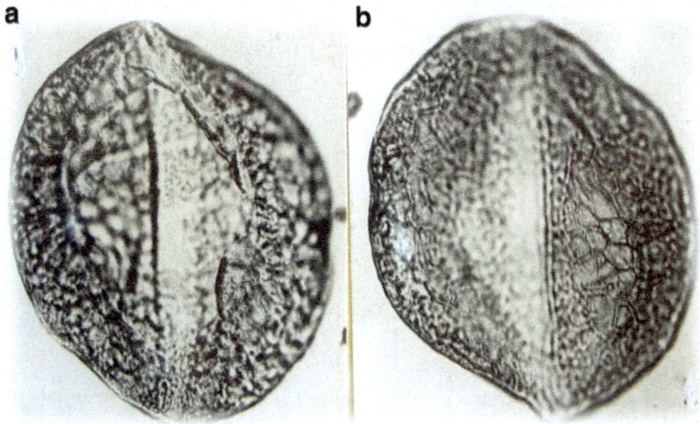

Fig. 2.61 **a** Distal face and 'sulcus'. **b** Half distal view (**a** and **b**) sulcus clear, taeniae faint possess micro-punctate structure. Ektexine intra-reticulate on sacci

Protohaploxypinus goraiensis (Potonie and Lele 1961; Hart 1964) (Fig. 2.61).
Occurrence: Upper Permian of Collie–Muja Basin. Frequent in the following samples: D/201-212″, B/ 224-227′.
Description: Disaccate striatiti. Diploxylonoid sacci, often slightly smaller than corpus. Corpus outline vaguely discernible. Cappa traversed by 4–6 striae bordering on wide taeniae, 8–16 μ wide and continuous, although not quite parallel. Ektexine detached equatorially on proximal face and sub-equatorially on distal face. Cappula sharply outlined, narrow, 5–10 μ wide. Nexine very thin, does not fold under distal roots. Ektexine thicker than nexine, intra-punctate on attached parts and intra-reticulate on sacci.
Dimensions: Total width: 70–100 μ; corpus diameter 50–84 (15 specimens measured).

Remarks and comparison: *P. goraiensis* is distinguished by the very thin, almost hyaline nexine, and sacci being smaller than corpus; taeniae wide and finely structured.

Genus ***Striatoabietites*** (Sedova 1956; Hart 1964).
Diagnosis: Disaccate striatiti. Diploxylonoid. Sacci significantly smaller than corpus. Corpus circular to sub-circular in lateral elongation. Cappa striated. Striae bordering on numerous narrow taeniae. Cappula wide. Structure of sacci intra-punctate to intra-verrucate.
Known distribution: Permian of the Northern and Southern hemispheres.
Discussion: *Striatoabietites* as emended by (Hart 1964) could include forms with rudimentary sacci such as *Striatoabietites multistriatus* Balme and (Hennelly 1955) and *Vittatina hiltonensis* Chaloner and Clark (1962). *Thuringisaccus* (Madler 1964) from the Triassic of Germany closely resembles *Striatoabietites* and could probably be treated as its synonym.
Striatoabietites was recorded and described by Backhouse (1991) from the Collie Basin, and from the Permian of Eastern India (Banerjee and D'Rozario 1990).

Striatoabietites multistriatus (Balme and Hennelly) (Hart 1964) (Fig. 2.62).
Occurrence: Sakmarian to Upper Permian of Collie–Basin. Artinskian to Upper Permian of Muja sub-Basin and Alexandra Bridge bore.
Description: Sub-saccate striatiti. Diploxylonoid or haploxylonoid. More commonly diploxylonoid. Sacci always distinctly smaller than corpus; often only ¼ corpus size. Corpus sub-circular to oval in lateral elongation. Cappa striated; striae consists of narrow taeniae, numbering 12–20, about 2 μ wide. Ektexine attached equatorially to slightly sub-equatorially on proximal face and sub-equatorially on distal face to form sub-sacci. Structure of attached as well as detached ektexine is intra-punctate to intra-granulate. Nexine usually thicker than ektexine, tending to fold under distal roots. Cappula relatively wide, from 1/3 to 2/3 of corpus width.
Dimensions: Total width: 45–65 μ; corpus: 30–45 μ (18 specimens measured).
Remarks and comparison: *Striatoabietites multistriatus* is distinguished by its numerous proximal taeniae/striae ad its pseudo-sacci which are devoid of recognisable structure. The species was recorded from Collie Basin by Backhouse (1991).

Genus ***Vittatina*** Luber ex (Wilson 1962).
Diagnosis: Asaccate striatiti. Bilaterally symmetrical. Proximal face taeniate; taeniae parallel to the long axis of the grain. Distal face smooth or possessing taeniae perpendicular to the long axis of the grain. Distal tenuitas analogues to the cappula in saccate pollen and is a characteristic feature.
Known distribution: Permian of Northern and Southern hemispheres.
Discussion: *Vittatina* was first described by Luber (1940, unpublished manuscript). Samoilovich (1953) described *Vittatina* forms, but he did not name a type species. Wilson (1962) designated *Vittatina subsaccata* as type species. *Striatolubrae* was erected by Hart (1962) and was later placed by him in synonymy with *Vittatina*.

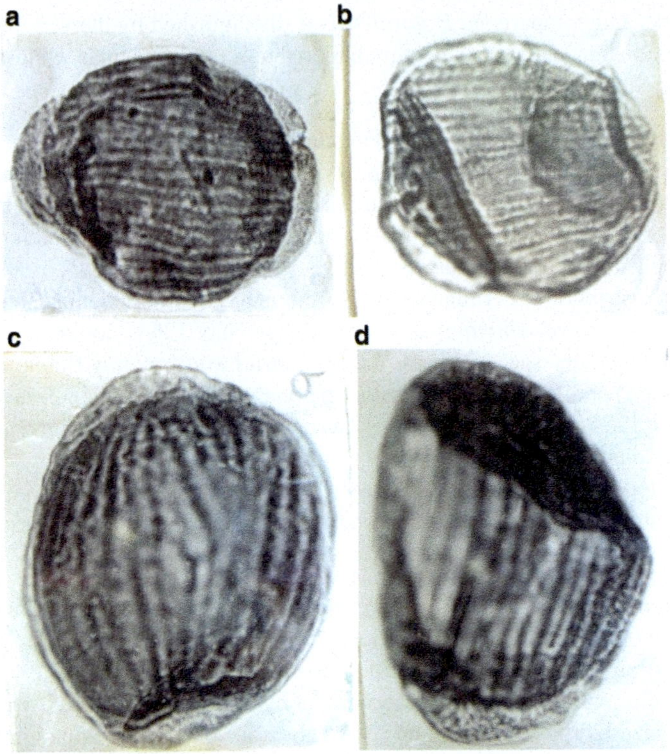

Fig. 2.62 **a** Distal view, **b** proximal view, **c** laesurae faintly visible from distal side. **d** Sulcus area with folds alongside

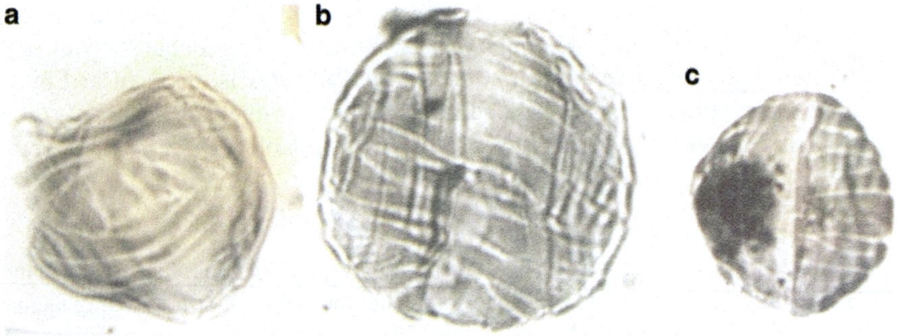

Fig. 2.63 **a** Distorted grain. Colpus varies in width (**b**, **c**)

Vittatina lucifer (Bharadwaj and Salujha 1964) (Fig. 2.63).
Occurrence: Uncommon; restricted to the Upper Permian of Collie–Muja Basin, Alexandra Bridge bore and Sue No. 1 in Perth Basin, where it is not found below the very top of the Upper Permian.

Fig. 2.64 Distal face

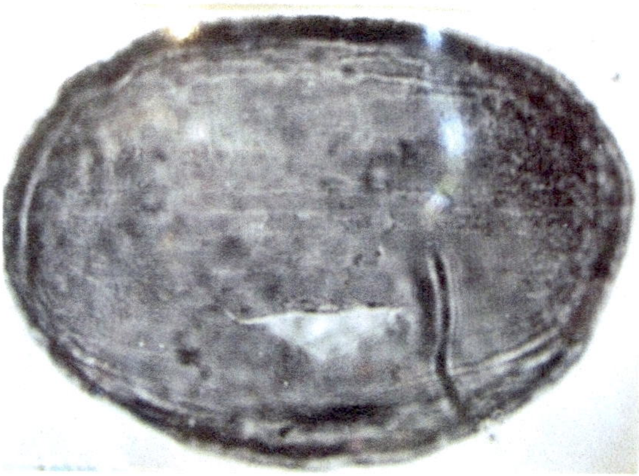

Description: A very small pollen grain, usually between 20 and 30 μ. Asaccate, taeniate. Outline sub-circular to oval. Both, proximal and distal face taeniate. Proximal taeniae number 10 or more. The peripheral taeniae are parallel to long axis of grain, while central taeniae are enclosed ovals. Distal taeniae number 2–3 and are perpendicular to proximal taeniae, continue to distal face following a 90° bend. Distal taeniae cover 2/3 of the pollen grain length, leaving a tenuinexinous non-sculptured cappula between them.
Dimensions: 21–33 μ across (12 specimens measured).
Remarks and Comparison: *Vittatina lucifer* was first described by Bharadwaj and Salujha (1964) from the Upper Permian of India. *Vittatina africana* described by Hart (1966) from the Lower Permian of South Africa closely resembles *V. lucifer*.

Vittatina sp (Fig. 2.64).
Occurrence: Uncommon. Lower to Upper Permian of Collie–Muja basin.
Description: Amb oval. Asaccate striatiti. Proximal face taeniate. Taeniae transversal, parallel to long axis of grain and run entire length of proximal face. Taeniae do not extend on to distal face which has very thin, hyaline ektexine. Due to the thin nature of the distal ektexine, ghost taeniae from the proximal face are seen through. Ektexine significantly thicker than endexine. Structure of ektexine is very fine intra-punctate on distal face and intra-granulate on proximal face. Protrusions of grana are visible over the outline of the grain.
Dimensions: Length: 75–90 μ, width: 30–75 μ (10 specimens measured).
Remarks and comparison: Vittatina sp. closely resembles *Tiwarisporites simplex* (Tiwari) Maheshwari and Kar (1967) recorded and illustrated by Backhouse (1991) from Collie Basin. Backhouse described the form as having taeniae/striae on distal as well as on proximal face. The form described here is striated on proximal face only which distinguishes it from *T. simplex*.

Vittatina scutata (Balme and Hennelly) (Bharadwaj 1962) (Fig. 2.65).
Occurrence: Artinskian to Upper Permian of Collie–Muja Basin.

Fig. 2.65 Distal face, sacci roots folded

Amb oval to sub-circular. Asaccate striatiti. Oroximal face striated. Taeniae numerous, about 15 or more extending on to distal face to about half way to centre of grain, covering about 1/3 of the pollen grain length on each side of the cappula. Nexine thin, barely detectable, folds along cappula. Structure of ektexine intra-granulate to intra-verrucate.
Dimensions: Long axis: 50–90 μ (10 specimens measured).
Remarks: A form which very closely resembles *V. scutata* described above has been recorded and described by Backhouse (1991) under *Tiwariasporites simplex* (Maheshwari and Kar1967). *V. scutata*-type pollen has also been recorded from the Permian of S. Africa (Anderson 1977) and Permian of Australia (Foster 1979) under different names.

Turma: **Plicates** (Naumova 1937) (Potonie 1960).
The PLICATES are sub-divided into three sub-turmae: 1.The PREACOLPATES (Potonie and Kremp 1955) to accommodate monosulcate pollen grains with a tetrad mark: 2. The POLYPLICATES (Erdtman 1955) to accommodate pollen grains with more than one groove or furrow running parallel to the long axis of the grain.
Chasmapollenites is described in the present study under sub-turma POLYPLICATES. It is justified by the fact that these pollen grains belong to a definite group of gymnosperms, namely the Gnetales.
Monocolpate grains such as *Cycadopites* (Wodehouse, Wilson and Webster 1946) are described in the present study under MONOCOLPATES (Iversen and Troels-Smith 1950). Hart (1965a, b) described Vittatina under group STRIATITI noting the difference though, namely that this genus is distinguished by the absence of sacci or the presence of only rudimentary sacci. It is emphasised

here that striate/taeniate pollen without sacci were found to be immature pollen of Glossopteris!

Sub-turma: PRAECOLPATES (Potonie and Kremp 1954).
Genus ***Marsupipollenites*** (Balme and Hennelly) (Balme 1970).
Diagnosis: Monocolpate, trilete. Amb of unexpanded grain sub-circular to oval. Expanded grain circular to sub-circular. Endexine thin, hyaline and smooth. Ektexine structured.
Known distribution: Permian of Australia (Balme and Hennelly 1956), Permian of India (Banerjee and D'Rozario 1990), Permian of Antarctica (Farabee et al. 1990).
Remarks and discussion: *Marsupipollenites* was originally described by Balme and Hennelly (1956) from the Permian of New South Wales. However, the form was recorded and described earlier from the Permian of Collie Basin, Western Australia, but not named (Balme 1952; de Jersey 1962; Dulhunty 1946).
Spores closely resembling *Marsupipollenites* were isolated by Pant and Nautiyal (1960) from the probable Pteridosperm sporangia *Polytheca elongata* (Pant and Nautiyal 1960). The close resemblance and possible affinity between *P. elongata* and *Marsupipollenites* was discussed by Pant and Nautiyal.

Marsupipollenites triradiatus (Balme and Hennelly 1956) (Fig. 2.66).
Occurrence: Abundant in all Upper Permian samples examined. Rare in the Artinskian and absent from Sakmarian samples.
Description: Monocolpate, trilete. Amb of unexpanded grain oval or oblate. Expanded grain circular. Proximally trilete; laesurae short and clearly outlined. Sutures occasionaly unequally developed. Distally monocolpate; colpus extend whole or almost whole length of grain, varies in shape from being evenly wide all along or slightly contracted in polar area, or widening at the extremities. Occasionally colpus bordered by folds. Colpus is made of endexine only, lacking in ektexinal layer. Endexine smooth, hyaline and frequently folded. In expanded grains endexine ruptured, rolled and folded along rims. Ektexine about 1 μ and structured. Structure is intra-punctate, intra-granulate to intra-verrucate with closely packed elements.
Dimensions: Length; 40–70 μ, width; 30–65 μ (25 specimens measured).
Remarks and comparison: *M. triradiatus* is distinguished from *Marsupipollenites striatus* (Balme and Hennelly 1956) by the character of its structure; elements are arranged in bands in *M. striatus*. There seems to be also a difference in its distribution; *M. striatus* being common in the Artinskian, while *M. triradiatus* in abundant in the Upper Permian.
Marsupipollenites reconstructed

Sub-turma: MONOCOLPATES (Wodehouse 1935; Iversen and Troel-Smith 1950).
Genus ***Entylissa*** (Naumova) (Potonie and Kremp 1955).
Diagnosis: Spindle-shaped, monocolpate pollen grain. Colpus extends whole length of grain, and widens at its extremities.

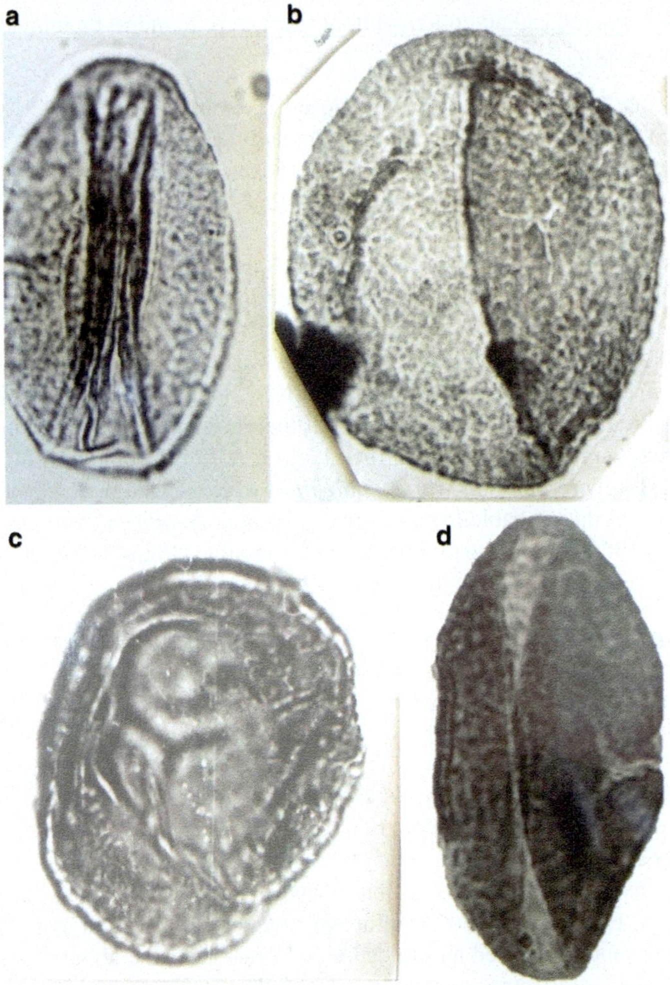

Fig. 2.66 **a** 'folded over' specimen, emphasising the monocolpate arrangement. **b** Semi-'unfolded' specimen. **c** Colpus extends almost entire width of pollen. **d** Monocolpate nature clear

Known distribution: Carboniferous of Russia (Luber 1938), Permian of New South Wales (Balme and Hennelly 1956), Permian of Collie, Western Australia (Backhouse 1991).

Discussion: The name *Entylissa* was given by Naumova (1937) to a form without naming a type. Potonie and Kremp (1955) named *Entylissa caperata* (Luber) as type species. *Entylissa* was emended by Leschik (1955) to include forms with horizontal striations. However, as Potonie and Kremp do not mention any particular exine properties in their diagnosis, the genus can include structured or low-relief sculptured forms as well, and Leschik's emending of the genus is superfluous.

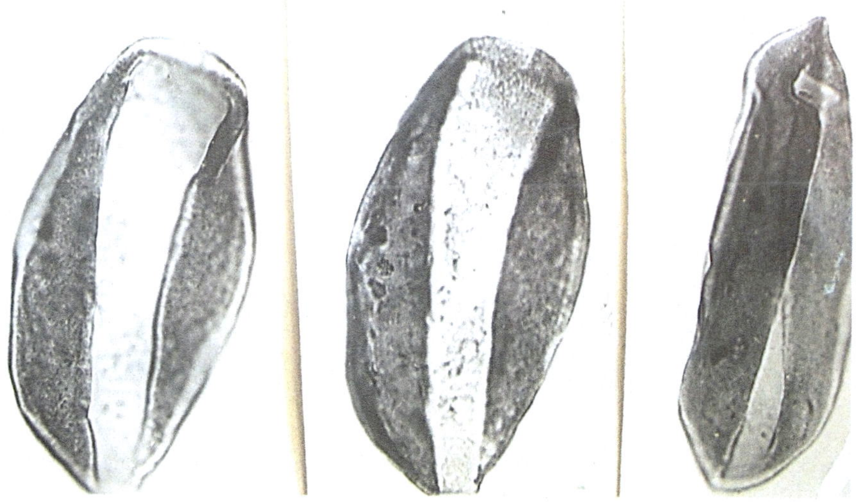

Fig. 2.67 *Entylissa cimbatus*

Entylissa cymbatus (Balme and Hennelly 1956) (Fig. 2.67).
Occurrence: Sakmarian tillite deposits, Collie-Muja Basin, Sue No. 1 bore in Southern Perth Basin.
Description: Monocolpate. Amb oval to fusiform. Colpus extends whole length of grain, is 9 μ wide in expanded specimens. Often widens slightly at its ends. In non-expanded grains colpus is a narrow slit bordered by longitudinal folds on each side. Exine is about 2–3 μ thick and of intra-punctate structure.
Dimensions: Length: 45–60 μ long; width: 21–31 μ (10 specimens measured).
Remarks and comparison: *Entylissa cymbatus* has been recorded and illustrated from Collie Basin (Backhouse 1991) under *Cycadopites cymbatus*. The two designated names are synonymous. *Entylissa cymbatus* has priority over *Cycadopites cymbatus*.

Sub-turma: POLYPLICATES (Erdtman 1952).
Genus ***Gnetaceaepollenites*** (Bharadwaj 1962).
Diagnosis: Ellipsoidal, inaperturate. Longitudinal grooves and accompanying ridges running whole length of spore are a distinct feature. Ektexine structure varies from fine intra-punctate to intra-granulate.
Known distribution: Permian of New South Wales (Balme and Hennelly 1952), Upper Permian of Collie (Balme 1952; Backhouse 1991) Permian of Karoo (Anderson (1977), Permian of Tasmania (Dulhunty and Dulhunty 1949), Upper Permian of India (Bharadwaj 1962), Upper Permian of Antarctica (Balme and Playford 1967). Polyplicate ephedroid pollen has been documented from Lower Permian to recent with great abundance in mid-Cretaceous (Hesse et al. 2000).
Discussion: The polyplicate character of this form distinguishes it from PRAECOLPATES. Although the Permian pollen are intensely flattened, it is still

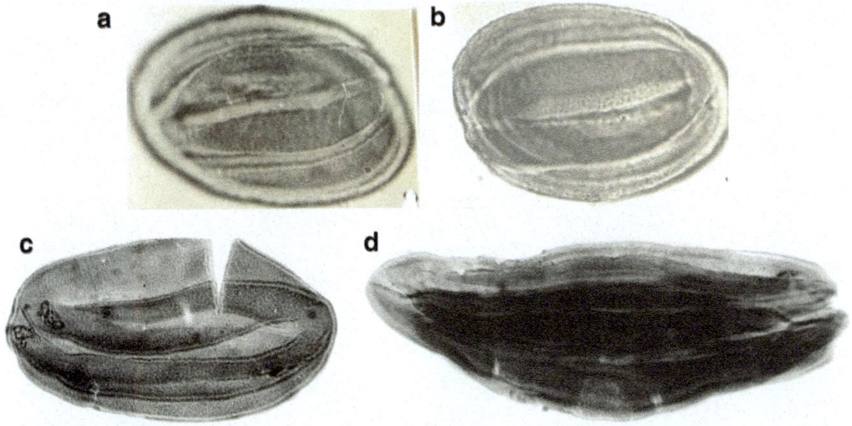

Fig. 2.68 **a** and **b** immature pollen *Gnetaceaepollenites*, **c** mature pollen of: Permian of Collie–Muja Basin. **d** Mature pollen (compressed and folded)

possible to 'unravel' them and reconstruct the specimens. The very close resemblance of *Praecolpites* to pollen of the gymnospermous **Gnetales** is interesting and no doubt was the reason that made Thiergart name *'Praecolpites'*-like forms *Gnetaceaepollenites*. Bharadwaj (1962) also recognised the resemblance to Gnetales and used the name ***Gnetaceaepollenites*** and ***Welwitschiapites*** (Bolchowitina 1953).

Praecolpites (Bharadwaj and Srivastava 1969) is considered in the present study as the synonym of *Gnetaceaepollenites* which is preferred by the author of this book because of the affinity to the living genus. The close resemblance to *Ephedra* pollen should not be ignored. The Gnetales are an old order of plants with remains in the geological record.

Gnetaceaepollenites sinuosus (Balme and Hennelly 1956; Bharadwaj 1962) (Fig. 2.68).

Occurrence: Permian of Collie–Muja Basin.

Description: Shape is oval-elongated. The variation in size is significant. The smaller specimens can be clearly described since they are not distorted by folding. The ribs vary in length but do not extend beyond polar region. Ribs are consistent in size, about 3 μ wide, divided by grooves. Endexine smooth while ektexine structured, intra-punctate to intra-granulate to microreticulate. Number of ribs 5–6.

Dimensions: Long axis: 55–150 μ; width: 35–90 μ (65 specimens measured).

Discussion: The variation in size and morphology within one species of Ephedra has been studied in detail (Ickert-Bond et al. 2003), and it has been shown to be misleading in assigning species on the basis of variation in pollen size, number of ridges, patterns of the grooves. Great variations in size were reported in an *Ephedra* species isolated from one and the same sporangium. Likewise, marked dimorphism in the pattern of the ridges was found in pollen from the same sporangium. The main difference between the Permian specimens and recent *Ephedra* seems to be by the number of ribs and accompanying grooves, which so far appear to be more numerous in recent forms than in the fossil ones. However, more specimen need to be studied in detail to confirm the difference.

Gnetaceaepollenites sinuosus was recorded and described from the Collie Basin by Backhouse (1991) under *Praecolpites sinuosus* and *P. ovatus.*

References

Anderson JM (1977) The biostratigtraphy of the Permian and Triassic part 3. A review of Gondwanan Permian palynology with particular reference to the Northern Karoo Basin, South Africa. Mem Bot Surv South Africa 41:1–67

Andrews HN, Mamay SH (1951) A new American species of Bowmanites. Bot Gaz 113:158–165

Backhouse J (1988) Trilete spores from the Collie Basin, Western Australia. Assoc Australasian Paleontol Mem 5:53–72

Backhouse J (1991) Permian palynostratigraphy of the Collie Basin, Western Australia. Rev Palaeobot Palynol 67:314–337

Balme BE (1959) Some palynological evidence bearing on the development of the Glossopteris flora. Symp Roy Soc Victoria 25:269–280

Balme BE (1952) The principal microspores of the Permian of Collie. In: Lord JH. Collie mineral field. West Aust Geol Surv Bull 105(2):164–201

Balme BE, Segroves KL (1966) *Peltacystia* gen.nov.; a microfossil of uncertain affinities from the Permian of Western Australia. R Soc West Aust J 19:26–31

Balme BE (1963) Plant microfossils from the lower triassic of Western Australia. Palaeontology 6(1):12–40

Balme BE (1969) The Permian–Triassic boundary in Australia. Spec Publ Geol Soc Aust 2:99–112

Balme BE (1970) Palynology of Permian-Triassic strata in the Salt Range and Surghar Range, West Pakistan. In: Kummel B, Teichert C (eds) Startigraphic boundary problems: Permian and Triassic of West Pakistan. University Press of Kansas, Dept. of Geology Special Publ, vol 4, pp 306–453

Balme BE, Hennelly JPF (1955) Bisaccate sporomorphs from Australian Permian coals. Aust J Bot 3(1):89–98

Balme BE, Playford G (1967) Late Permian plant microfossils from the Prince Charles Mountains, Antarctica. Rev Micropal 10:179–192

Balme BE, Hennelly JPF (1956) Monolete, Monocolpate, and Alete Dporomorphs from Australian Permian coals. Aut J Bot 4(1):54–67

Balme BE, Hassell CW (1962) Upper Devonian spore from the Canning Basin, Western Australia. Micropalaeontology 8(1):1–28

Banerjee M, D'Rozario A (1990) Palynostratigraphic correlation of Lower Gondwana sediments in the Chuparbhita and Hura Basins. Rajmahal Hills, Eastern India

Bharadwaj DC (1962) The miospore genera in the coals of Raniganj Stage (Upper Permian), India. The Palaeobotanist 9:68–106

Bharadwaj DC (1964a) Potonieisporites Bharadwaj: Ihre Morphologie, Systematik und Stratigraphie. Fortsch Geol Rheinld Westf 12:45–54

Bharadwaj DC (1964b) The organisation in pollen grains of some early conifers. The Palaeobotanist 12:18–27

Bharadwaj DC, Salujha SK (1964) Sporological study of Seam VIII in Raniganj coalfield, Bihar (India), Part I; Description of sporae dispersae. The Palaeobotanist 12:181–215
Bharadwaj DC, Tiwari RS (1964) On two monosaccate genera from Barakar Stage of India. The Palaeobotanist 12:139–146
Bharadwaj DC, Srivastava SC (1969) A Triassic microflora from India. Palaeontographica 125:119–149
Blackburn KB (1936) A green alga. Trans R Soc Edin 58:841
Cazzulo-Klepzig M, Mendonca Filho JG, Guerra-Sommer M, Menezes T, Wagner-Simas M, Mendonca JO, Degani-Schmidt I (2009) Effect of volcanic ash-fall on Permian peat—forming environment, on the basis of palynology, palynofacies and palaeobotany (Faxinal Coalfield, Brazil). Rev Bras Palentol 12(3):179–194
Chaloner WG, Clarke RFA (1962) A new British Permian spore. Paleontology 4(4):648–652
Churchill DM (1960) Living and fossil unicellular algae and aplanospores. Nature 186:493–497
Clarke RFA (1965) British Permian saccate and monosulcate mispores. Paleontology 8(2):322–354
Cookson IC (1947) Plant microfossils from the lignites of the Kerguelen Archipelago. B.A.N.Z. Antarct Exped (A) 2(8):129–142
Cookson IC, Dettmann ME (1959) On Schizosporis, a new form genus from Australian Cretaceous deposits. Micropaleontology 5(2):213–216
Cookson IC, Eisenack A (1962) Some Cretaceous and Tertiary microfossils from Western Australia. Proc R Soc Victoria 75(2):269–273
Cousimer HL (1965) Permian spores from Apillapampa, Bolivia. J Palent 39(6):1097–1112
Deflandre G (1954) Systematique de hystrichosphaerides, sur laception de genre Cymatiosphaera Wetzel. C.R. Somm Soc Geol France 1954:257–258
de Jersey NJ (1962) Triassic spores and pollen grains from the Ipswich Coalfield. Geol Surv Queensl Publ 307:1–16
Dettmann ME (1963) Upper Mesozoic microfloras from southeastern Australia. R Soc Victoria Proc 77(1):148 pp
Downie C, Sarjeant WAS (1963) On the interpretation and status of some hystrichosphere genera. Palentology 6(1):83–96
Dulhunty JA (1946) Principal microspore types in the Permian coals of New South Wales. Proc Linn Soc N.S.W. 70:132–139
Dulhunty JA, Dulhunty R (1949) Notes on microspore types in Tasmanian Permian coals. Proc Linn Soc N.S.W. 74:132–139
Eisenack A (1962) Neue problematische mikrofossilien. N Jb Geol Palaeont Abh 114(3):260–265
Erdtman G (1957) Pollen and spore morphology, plant taxonomy, gymnosperms. Almquist & Wiksel, Stockholm, 151 pp
Erdtman G (1952) Pollen morphology and plant taxonomy, I angiosperms, Uppsala, 539 pp
Evans PR (1964) Lower Permian microflora from the Crown Point Formation, Finke area, Northern Territory. B.M.R. Records 196(30)
Evitt WR (1963) A discussion and proposals concerning fossil dinoflagellates, hystrichospheres and acritarchs. Proc Nat Acad Sci 49(3):298–302
Farabee M, Taylor EL, Taylor TN (1990) Correlation of Permian and Triassic palynomorph assemblage from central Transantarctic Mountains, Antarctica. Rev Palaeobot Palynol 65:257–265
Foster CB (1979) Permian plant microfossils of the Blair Athol Coal Measures, Baralaba Coal Measures and basal Rewan Formation of Queensland. Queensl Geol Surv Publ 372 Palaeontol Pap 45:1–244
Grebe H (1957) Zur Microflora des Niederrheinischen Zechsteins. Geol Jb 73:51–74
Hart GF (1960) Microfloral investigation of the Lower Coal Measures (K2) Ketewaka-mchuchuma Coalfield, Tanganyika. Geol Surv Tanganyika Bull 30:(1–18)
Hart GF (1963) A probable pre-Glossopteris microfloral assemblage from Lower Karroo sediments. S Afric J Sci 59(5):135–46
Hart GF (1964) A review of the classification and distribution of the Permian mispore taxon. Disaccate Striatiti: C.R. 5th Cong Int Strat Carb Paris 49

Hart GF (1965a) Microflora from the Ketewaka-Mchuchuma Coalfield, Tanganyika. Geol Surv Tanganyika Bull (36):1–25
Hart GF (1965b) The systematics and distribution of Permian miospores. Wiwatersrand Uni. Press, Johannesburg, 252 pp
Hart GF (1966) Vittatina Africana, a new miospore from the Lower Permian of South Africa. Micropaleontology 12(1):37–42
Hemer DO, Nygreen PW (1967) Algae, acritarchs and other microfosils from Lower Carboniferous of Saudi Arabia. Micropaleontology 13(2):183–194
Hesse M, Weber M, Halbritter H (2000) A comparative study of the polyplicate pollen types in Arales, Laurales, Zinjiberales amd Gnetales. Morphol Biol 227–239
Höeg OA, Bose MN (1960) The Glossopteris flora of the Belgian Congo. Ann Mus R Congo Belg Tervuren 8(32):106
Ibrahim AC (1933) Sporenformen dess Agir horizont des Ruhr reviers. Diss.T.H. Berlin, pp 1–46
Ickert-Bond SM, Skvarla JJ, Chissoe WF (2003) Pollen dimorphism in Ephedra L. (Ephedraceae). Rev Palaeobot Palynol 124:(325–334)
Iversen J, Troels-Smith J (1950) Pollenmorfologiske definitioner og typer. Damm Geol Unders IV 3(8)
Jansonius J (1962) Palynology of Permian and Triassic sediments, Peace River Area, Western Canada. Palaeontographica 110B:1–18
Jizba KMM (1962) Late Paleozoic bisaccate pollen from the United States midcontinent area. J Paleont 36:871–887
Klaus W (1963) Sporen aus dem sudalpinen Perm. Jb Geol BA 106:229–363
Knox EM (1950) The spores of Lycopodium, Phylloglossum, Selaginella and Isoetes and their value in the study of microfossils of Palaeozoic age. Trans Proc Bot Soc Edinburgh 35:209–257
Kosanke RM (1950) Pennsylvaniamn spores of Illinois and their use in correlation. Ill. State Geol/Surv Bull 74:1–125
Lakhanpal RN, Sah SCD, Dube SN (1958) Further observations on plant microfossils from a carbonaceous shale near Naini Tal. The Palaeobotanist F 1(2):111–120
Lele KM, Maithy PK (1964) An unusual monosaccate spore from the Karharbar stage, Gindih Coalfield, India. The Palaeobotanist 12(3):307–312
Leschik G (1955) Upper sporen aus dem Salzton des Zechsteins von Neuhof bei Fulda. Palaeontographica 100B:122–142
Leschik G (1956) Sporen aus dem Karru-Sandsteine von Norronaub (sudwest Africa). Senck leth 40(1/2):51–95
Li J, Changqun C, Servais T, d'Ascq V, Zhu J (2004) Later Permian acritarchs from Meishan (SE China) in context of Permian palaeontology and Palaeoecology
Lindström S, McLoughlin S, Drinnan AN (1997) Intraspecific variation of taeniate bisaccate pollen within Permian Glossopterid sporangia, from the Prince Charles Mountains, Antarctica. Int J Plant Sci 158(5):673–684
Machiavello-Martinez JC (1963) Microsporomorfes tipos contenidos en el glacial de la base delasistema de Gondwana del Uruguay. Bot da Univers d Parana 10(1):1–14
Mcloughlin S (1992) Western Australia's Permian plants. Fossil Collect Bull 38:25–31
Madler K (1964) Die geologische Verbreitung von sporen und pollen in der Deutschen Trias. Beih Geol Jb 65:147
Medvedeva (1960) In: Zhou YX (2019) Rev Palaeobot Palynol 81(2–3):193–211
Mehta KR (1944) Microfossils from a carbonaceous shale from the Pali beds of South Rewa, Gondwana Basin. Proc Nat Acad Sci India 14(4/5):125–141
Maheshwari HK (1967) Studies of the Glossopteris flora of India—29 mispore assemblage from the Lower Gondwana exposures along Bansloi River in Rjmahal Hills, Bihar. The Palaeobotanist 15(3):258–280
Mishra S, Jha N (2017) Early Permian (Asselian-Sakmarian) palynoflora from Chintalapudi Area, Godavari Graben, South India and its palaeoenvironmental implications. J Palaentol Soc India 62(2):23–40

Naumova SN (1953) Pollen of angiosperm type from Lower Carboniferous deposits. Izv Akad Nansk S.S.S.R Geol Ser 3:103–113
Naumova SN (1961) Spore and pollen complexes of the Carboniferous on the Russian platform and in the Urals. CR Congr Geol et Str du Carb Pt 2:437–442
Nilsson T (1958) Uber des vorkommen eines mesozoischen sapro-pelgesteinge in Schonene. Publ Inst Miner Paleont Quat Geol Uni Lund 53:1–111
Norris G (1965) Triassic and Jurassic miospores and acritarchs from the Beacon and Ferrar groups, Victoria Land, Antarctica. NZ Geol Geophys 8(2):236–277
Norris G, Sarjeant WA (1965) A description index of fossil Dinophyceae and Acritarchs, Wellington. NZ Geol Surv Palaeontol Bull 40:71 pp
Nygreen PW, Bourn OB (1967) Morphological variation of *Potonieiesporites* in late Pennsylvanian Florule. Rev Palaeobot Palynol 3(14):325–332
Pant DD (1954) Suggestions for the classification and nomenclature of fossil spores and pollen grains. Bot Rev 20:33–60
Pant DD (1955) On new disaccate spores from the Bacchus Marsh tillite, Victoria, Australia. Ann Mag Nat Hist Ser 12(8):41–64
Pant DD, Nautiyal DD (1960) Some seeds and sporangia of Glossopteris Flora from Raniganj Coalfield, India. Palaeontographica 107:41–64
Pant DD, Mehra S (1963) On the occurrence of Glossopterid spores in the Bacchus Marsh tillite, Victoria, Australia. Grana Palyn 4(1):111–120
Playford G (1965) Plant microfossils from the Triassic sediments near Poatina, Tasmania. J Geol Soc Aust 12(2):173–210
Potonie R (1956) Synopsis der Gattungen der sporae dispersae. Teil 1. Beih Z Geol Jahrb 23:244 pp
Potonie R (1958) (1960) Synopsis der Gattungen der Sporae dispersae teil I-III. Beih Geol Jahrb 31, 39:1–189
Potonie R, Klaus W (1954) Einige Sporengattungen das alpiner salzgebirges. Geol Jb 68: 517–546
Potonie R, Kremp G (1954) Die gettungen der palaontologischen sporae dispersae und ihre stratigraphie. Geol Jb 69:111–194
Potonie R, Kremp G (1955) Die Sporae Dispersae des Ruhr Karbons. Teil I: Palaeoontographica 98B:1–136
Potonie R, Kremp G (1956) Die Sporae Dispersae des Ruhr Karbons. Teil II. Palaeontographica 99B:85–191
Potonie R, Schweitzer NJ (1960) Der Pollen von *Ulmania frumentaria*: Palaeont Z 34(1):27–39
Potonie R, Lele KM (1951) Studies in the Talchir flora-I, South Rewa, Gondwana Basin. The Palaeobotanist 8(1&2):22–37
Potonie R, Sah SCD (1960) Sporae dispersae of lignites from Cannhore Beach on the Malabar coast of India. The Paeobotanist 1(1):121–135
Richardson JB (1965) Middle Old Red Sandstone spore assemblages from the Orcadian Basin, North East Scotland. Paleontology 7(4):559–605
Rigby JF, Hekel H (1977) Palynology of Permian sequence in Springsure Anticline, Central Queensland. Publ Geol Surv Qld 363:37 pp
Samoilovich SR (1953) Pollen and spores from the Permian deposits of the Cherdyn and Aktyubinsk areas, Cis-Urals. Palaeobot Sbornik na 75:5–57
Schaarschmidt F (1963) Sporen und hystrichosphaeriden aus dem Zechstein von Sudingen in der Wetterau. Palaeontographica 113B:38–91
Schopf JW, Wilson LR, Bentall R (1944) An annotated synopsis of Palaeozoic fossil spores and the definition of generic groups Ill. Geol Surv Rep 91:1–73
Scott RA, Barghoorn ES, Leopold EB (1960) How old are the angiosperms. Am J Sci 258A:284–299
Sedova NA (1956) The definition of four genera of disaccate Striatiti. Translated from Russian G. Hart. Johannesburg
Segroves KL (1967) Cutinized microfossils of probable non-vascular origin from the Permian of Western Australia. Micropalaeontology 13(3):298–305
Singh HP (1964) Miospore assemblage from the Permian of Iraq. Paleontology 7(2):240–266

Staplin FL (1960) Upper Mississippian plant spores from the Golata formation, Alberta, Canada. Palaeontographica 107B:1–40

Tiwari RS (1964) New miospore genera in the coals of Barakar Stage (Lower Gondwana) of India. The Palaeobotanist 12(3):250–159

Tiwari RS (1965) Miospore assemblage in some coals of Barakar stage (Lower Gondwana) of India. The Palaeobotanist 12(2):168–214

Tiwari, Navale (1967) In: Yong, Servis, Quinglai (2013) The diversity in the Permian phytoplankton. Rev Palaeobot Palynol 198:145–161

Tripathi A, Vijaya, Murthy S, Chakarborty B, Das DK (2012) Stratigraphic status of coal horizon in Tatapan-Ramkola coalfield, Chhattisgarh, India. J Earth Syst Sci 121(2):537–557

Virkki C (1945) Spores from the Lower Gondwana of India and Australia. Proc Nat Acad Sci Ind 15(4–5):93–175

Wetzel O (1933) Die organischer Substanz erhalten Microfossilen. Palaeontographica Abt. A 77:147–148 & 78:1–110

Weyland H, Krieger W (1953) Die Sporen und Pollen der Aachene-Kreide und ihre Bedeutung fur die characterisierung des Mittleren Senons. Palaeontographica 95B:6–29

Wilson LR (1962) Plant microfossils from the flowerpot formation. Okl Geol Surv Circ 49:1050

Wilson LR, Webster RM (1946) Plant microfossils from a Fort Union Coal of Montana. Am J Bot 33(4):271–278

Wodehouse RP (1935) Pollen grains. Their structure, identification and significance in science and medicine. McGrow-Hill, New York, 74 pp

Zavialova NE, Gomankov A, Yaroshenko OP, Rovnina LV (2004) Morphology and ultrastructure of some monosaccate pollen grains *of Cordaitina* Samoilovich 1953 from the Permian of Russia. Acta Palaeobotanica 44(1):3–35

Chapter 3
The Microfloral Assemblages—Their Environmental and Climatic Interpretation

The questions arising from the systematic work of this study are: What do the pollen and spores tell us? How significant are the sometime subtle changes in pollen/spore assemblages? Is the composition of the spore/pollen assemblage indicative of environmental and atmospheric conditions? In order to answer these questions and reconstruct their environment, there is a need to organise the plant remains, namely the pollen, spores and acritarchs into some grouping with common characteristics. These are: appearing together in same stratigraphic level; sharing morphological features; abundance in a lithological unit with specific petrology; prevalence of one type of pollen, spore or acritarch.

On the basis of the above, the microfloral assemblages are grouped into four biostratigraphic units (Figs. 3.2, 3.4, A.1):

Biostratigraphic Unit I is characterised by the following pollen: *Parasaccites gondwanensis*, *Microbaculispora tentula* and *Punctatisporites gretensis*. These pollen comprise the oldest assemblage isolated and described from the Permian sequence of the Collie samples, and may be referred to as the *Parasaccites* Zone. This unit is present in all bore hole samples of Collie, Wilga Basins, Muja sub-basin and Sue No. 1 bore. The lithology that this assemblage shares is tillites, which are glacial deposits and contain the earliest Permian vegetation following the retreating glaciers. This assemblage is poor in types of pollen on the generic level; the main components of the microflora being the three pollen mentioned above. The ratios between the three forms vary slightly from one sample to another. Nevertheless, they comprise 80–90% of the total microflora of the *Parasaccites* zone. *Parasaccites* and *Punctatisporites gretensis* dominate this assemblage, although there are bisaccate forms present, namely *Potoniesporites neglectus* and *Illinites tectus*. The spores of the monolete *Cycadopites cymbatus* and the trilete *Microbaculispora tentula* accompany the assemblage. All these pollen except *Illinites* do not have any significant presence beyond the Sakmarian. Illinites continues to be present in the Artinskian Biostratigraphic Unit II.

Towards the end of the Sakmarian a change to less harsh climatic conditions enabled a group of new plants to establish themselves. These are represented by the spores of *Leiotriletes*, *Acanthotriletes*, *Retusotriletes*, *Apiculatisporites and*

M. Glikson-Simpson, *Coal—A Window to Past Climate and Vegetation*,
https://doi.org/10.1007/978-3-030-44472-3_3

Cristatisporites, and the pteridosperm *Marsupipollenites*, and a few disaccate pollen producing plants. The first appearance of Striatiti representing the Glossopteris flora appeared at the end of the Sakmarian. The spores which represent ferns and *Marsupipollenites*, a Pteridosperm, both groups form understory plants in the Cordaitalean swamp forest, with the odd Equisetum-like plant represented by the spores of Cycadopites cymbatus.

Other pollen types occurring in small numbers are as mentioned: occasional disaccate striatiti; the disaccate form *Illinites* and trilete spores such as *Apiculatisporites levis, Retusotriletes diversiformis* and the zonate spore *Cristatisporites.* Assemblage No. 1 corresponds To Balme's (1964) *Nuskoisporites* assemblage and Evan's (1967) Stage I. The same pollen/spore assemblage was recorded from the Nangetty Formation in Perth Basin, which was dated as Sakmarian (Playford 1959). On the basis of correlation between the samples dominated by the *Parasaccites* assemblage and the Nangetty formation, Sakmarian age is assigned to Unit I in Collie–Muja Basin.

Helby (1967) in discussing the Permo-Carboniferous boundary in Australia brings in the microfloral factor, namely the introduction of the '*Potoniesporites*' microflora following the decline of the 'Lycopsid' microflora, "an event which appears to be intimately associated with the widespread development of 'glacigene' sediments…" Helby regarded the *Potoniesporites* microflora as the top of the Carboniferous, "the closest, available approximation to, and most convenient location for the top of the Carboniferous system in Australia". *Potoniesporites* represented a certain vegetation which gave way to the *Parasaccites*-dominated plant community, which marks the Lower Permian. *Parasaccites*' a monosaccate distinctive pollen is typical of the **Cordaitales** dominating the Sakmarian tillites ('glacegene sediments').

Cordaitales plant remains are known from coal deposits in the Carboniferous and Permian of Western Europe, North America, North China and Russia. Fossil evidence indicates that these plants attained over 30 m in height (Cridland 1964) had slender, well-developed branches with spirally arranged, leathery leaves of parallel venation. Leaves attained a length of 1 m. Cridland (1964) reconstructed Codaitalean plants from fossil remains (Fig. 3.1).

In summary, the Sakmarian floral community can be seen as dominated by the cordaitales tall trees with the understory of Pteridosperms and ferns. Their abundance and preservation enable their identification and the reconstruction of the post-glacial plant scene.

In bore B, Collie Basin (Fig. 3.3)

In bore D, Collie Basin (Fig. 3.2)

In Sue-1 Bore, South Perth Basin (Fig. 3.4)

Chart of individual pollen species in Collie bore B (Appendix A.1 and A.2)

While *Parasaccites gondwanensis* is a dominant form in the tillite sediments, *Microbaculispora tentula* becomes more abundant at the top of the tillites and the overlying sandstone. *P. gondwanensis*, on the other hand, decreases in quantity towards the top of the tillites and the overlying sandstones, where it drops to about 10% of the total microfloral assemblage from 80% in the tillites. There is a link between *Parasaccites* and cold climate.

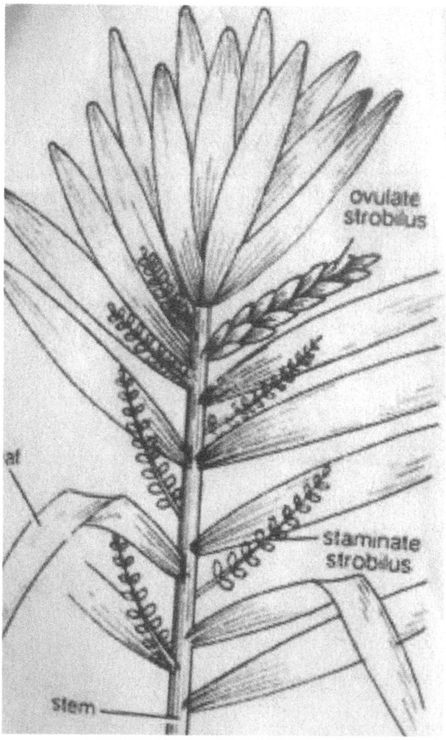

Fig. 3.1 Reconstructed Codaitalean plant (Cridland 1964)

High concentrations of monosaccates in the cold zone of the Lower Permian is also reported by Wheeler and Götz (2016) from the Highveld coal deposits of the Karoo Basin. Wheeler and Götz's study focuses specifically on the palaeoenvironment and palaeoclimate of early Permian. Likewise, Götz et al. (2017) record the same assemblage from the Sakmarian of Mozambique, and 'bio-zones' similar to the biostratigraphic units of the Collie Basin.

The equivalent of Biostratigraphic Unit I can be traced through all Gondwana Basins, representing the earliest vegetation of the tillites, the early Permian glacial deposits. The unit is marked by the pollen *Parasaccites gondwanensis* and is termed the ***Parasaccites* zone**. The same pollen grain, although often under different names is also reported from other Gondwana basins: Antarctica (Farabee et al. 1990); Eastern India (Banerjee and D'Rozario 1990); South India (Mishra and Jha 2017); Karoo Basin, South Africa (Anderson1977; Wheeler and Götz 2016; Götz et al. 2017).

The problem that emerges when trying to compare the pollen/spore assemblages of all the Gondwana Basins as they relate to units within the Permian system is the naming of the genera and species. When there are illustrations provided in many publications, then it is made possible. However, when illustrations are absent, unclear or only names are supplied, any attempt at correlation is impossible. It is then that the practice of assigning forms under different names becomes a fruitless and useless exercise.

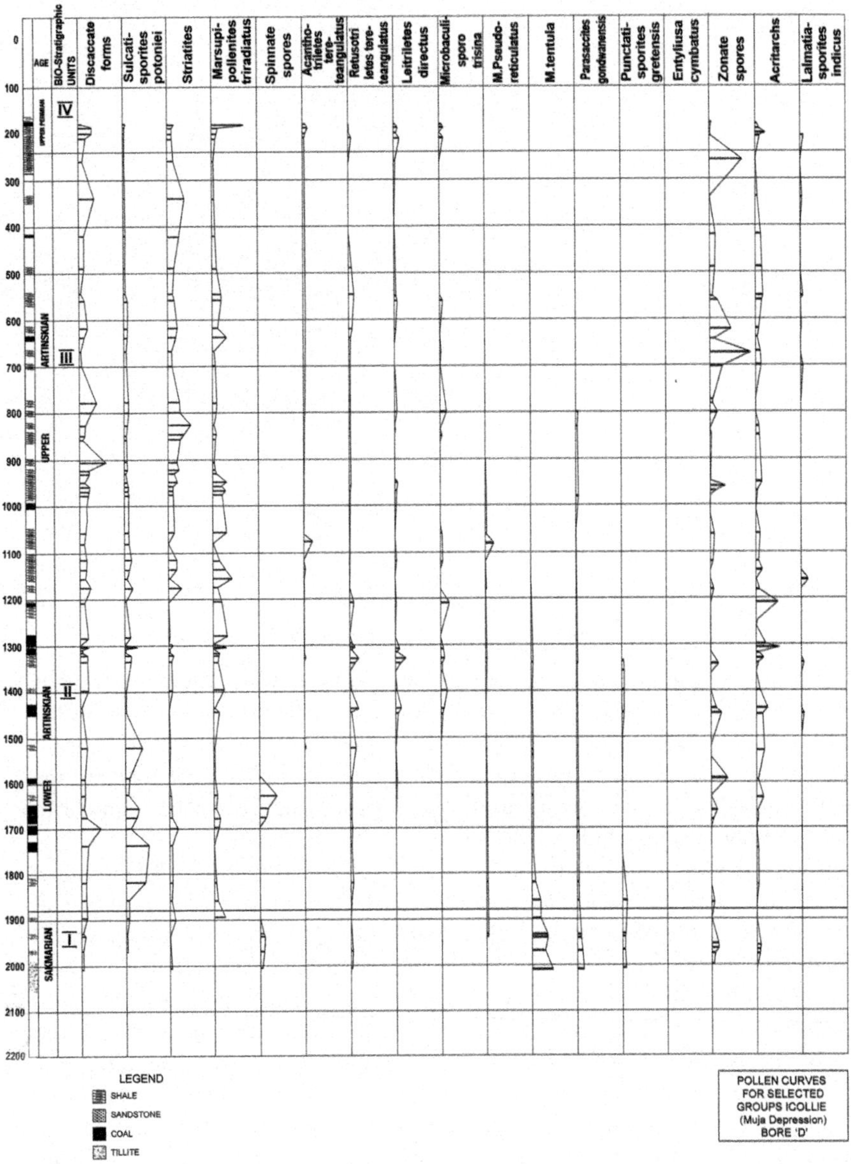

Fig. 3.2 Pollen curves for selected groups in Collie (Muja depression) SITE 'D' Bore

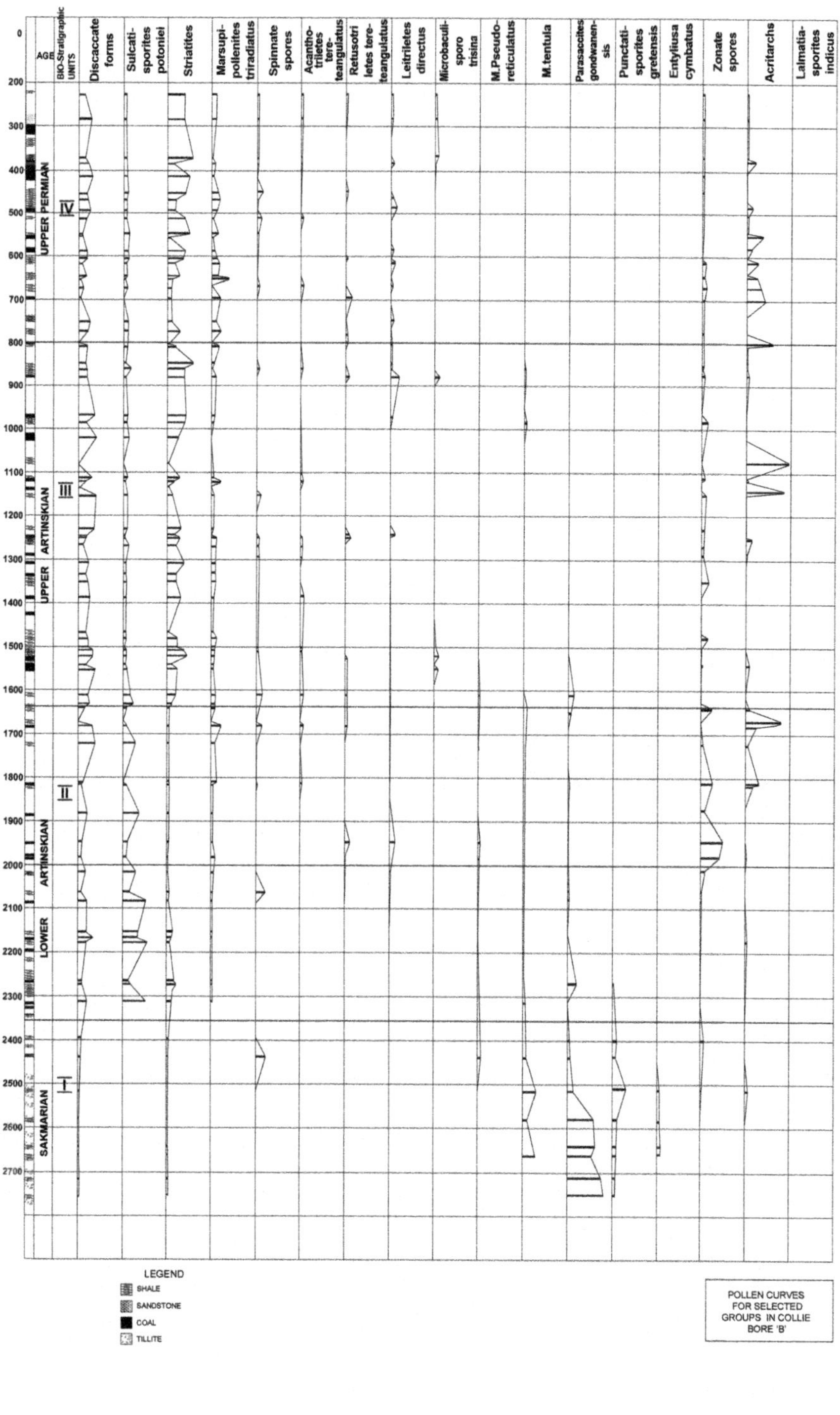

Fig. 3.3 Pollen curves for selected groups in Collie; SITE 'B' Bore

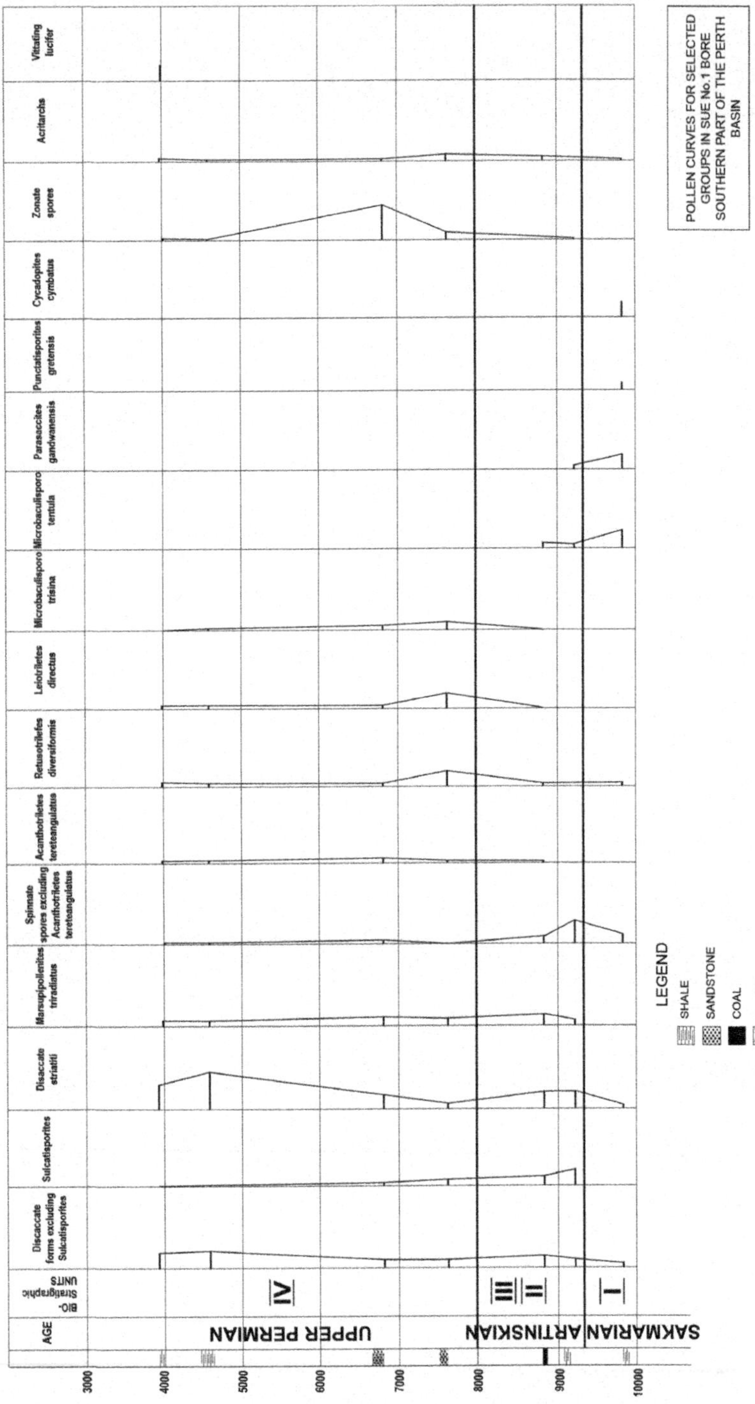

Fig. 3.4 Pollen curves for selected groups in Sue No.1 Bore; Southern Perth Basin

Biostratigraphic Unit II—The Lower Artinskian is characterised by the dominance of the saccate pollen *Sulcatisporites potoniei*, and accompanying spores; the trilete spinate *Acanthotriletes and Microbaculispora* are also conspicuous members of this assemblage. Unit II is referred to as ***Sulcatisporites* zone**. This assemblage is found in siltstones and sandstones overlying the tillites. The entire sequence of tillites and overlying sandstones and siltstones was referred to as the Stockton Formation (Low 1958). This sequence consists of coal measures, referred to as the Collie Formation or the Lower Coal Measures (Lord 1952). The Collie Coal Measures correspond to the Irwin River Coal Measures in the northern Perth Basin (Playford 1959). The latter have been dated as Artinskian (McWhae et al. 1958).

Microbaculispora tentula and *Parasaccites gondwanensis* dominant in the tillites of UNIT I decrease in numbers dramatically towards the end of the Sakmarian and have only a sporadic presence in Zone II; they are overrun by *Sulcatisporites potoniei* and *Acanthotriletes tereteangulatus*. Generally, the microflora in Unit II is richer in pollen and spore forms than Unit I. Disaccate pollen usually having a very small isolated appearance in UNIT I increase notably in UNIT II (Appendices B, C, D). *Marsupipollenites* is present (Figs. 3.2, 3.3 and 3.4).

Zonate spores such as *Kraeuselisporites* and *Gondispora* reach in Collie Basin 26%, smaller numbers in Sue No. 1 bore in the Perth Basin. Zonate spores are produced today by ferns such as *Lycopodium complanatum* and *Angiopteris* as well as by some species of *Selaginella* and *Asplenium*. They can thrive only in wet conditions. The Lower Artinskian therefore was wet and still cold.

The change in the composition of pollen and spore types in the ***Sulcatisporites* Zone** signifies a change in vegetation which follows a change in climatic conditions. The changes in the depositional environment as expressed in the vegetation community are recorded by the fossil pollen and spores. The plant community that existed in UNIT I dominated by the Cordaites and represented by *Parasaccites gondwanensis* became sparse. Likewise, *Punctatisporites gretensis* has only a trace presence in UNIT II. The Cordaites became extinct, and have no representatives in younger geological formations.

The 'lowlands' swamps with varying water level have been taken over by fern-like plants, forming the under story, including pollen of *Marsupipollenites* and other spores such as several species of *Microbaculispora* and spores of *Retusotriletes*. On the other hand, Calamites represented by monolete spores such as *Laevigatosporites* are horsetails, the major contributors to the coal deposits in the Northern hemisphere, are also represented in the Collie-Perth Basin, Permian and also in the equivalent zone of the Lower Gondwana sediments in India (Banarjee and D'Rozario 1990). They have today a sole representative in the herbaceous *Equisetum*. Zhuo Feng (2017) in his reconstruction of the Permian flora in China, based on fossil forests, describes the Calamites as "fast growing upright, forming bamboo-like thickets and an extensive prostrate rhizome system." The rhizome system is reminiscent of today's mangroves.

In the second half of the Lower Artinskian, water level in the wetlands had risen. A sign of a change in the climate towards higher rainfall, flooding of lowland and forming of lakes is evident from the peaks of acritarchs (Figs. 3.2 and 3.3). Peaks in acritarchs are met with extreme lows in disaccate pollen, including *Sulcatisporites* and close to disappearance of the Striatiti. As the swamps were flooded, the Glossopteris trees were drowned and there is no existence of them until a return to the swamp environment. The high rainfall/extreme wet was prone to fluctuations and the Glossopterids return is evident in the Upper Artinskian where they once again make up a significant contribution to the vegetation.

In the top half of the Upper Artinskian acritarchs peak twice accompanied by lows in Striatiti and disaccates, both the disaccates and Striatiti disappear entirely during the peak of acritarchs (Figs. 3.2, 3.3 and 3.4), signifying very high water level and a lacustrine depositional environment as a result of high rainfall and episodal climate change.

Biostratigraphic Unit III—The Upper Artinskian: This unit is marked by the significant fall in numbers of *Sulcatisporites* to almost its disappearance, and the marked increase in the **Striatiti**. Unit III corresponds approximately to Evans' (1967) Stage 4 and the Upper part of the *Vittatina* assemblage of Balme (1964).
The Striatiti are the pollen of the **Glossopterids** which appeared in Gondwana in the early Permian and became extinct in the Triassic. The Glossopteris flora was widespread in Gondwana during the Permian, and fossil plant remains and pollen were found in Australia, India, Antarctica, South America, South Africa and Madagascar. The most dramatic change of flora more obvious in the Upper Artinskian is the result of warming from very cold to cold temperate bringing on the explosion in the flora that produced the Striatiti, namely the Glossopteris Flora. The G. flora diversified to the extent of becoming the predominant vegetation of Gondwana. The significant number of genera and vast number of species is a testimony to the abundance of the Glossopterids.

The Glossopterids are generally assigned to the gymnosperms and some favour the order Gingkoales. The pollen, which is the ultimate generic feature that defines and distinguishes one plant from another plant (Walker et al. 1968; Guppy et al. 1973), supports the connection to the gymnosperms. The striate–taeniate feature is unique to the striatiti and has no match among living gymnosperms. Interestingly, there are other common features: The venation of leaves of some Glossopteris remains are strikingly similar to those of *Agathis australis*, others to *Podocarpus elatus*. So much so that if found fossilised leave, *Agathis* could be easily mistaken for *Glossopteris*.

The slight warming of Gondwana led to the formation of extensive areas of swamp which were conditions favouring the Glossopteris plant communities. Fossil root structure shows the special characteristics that support swamp forests similar to today's *Taxodium* swamp forests.

Although the Striatiti appear in small numbers in the Lower Artinskian, they come into their own in the Upper Artinskian and dominate the pollen assemblage. The diversification of fern species continued in the form of many species of zonate spores. *Marsupipollenites* is still a distinct contributor to the under story of the swamp forests. The Glossopteris forests were accompanied by a rich and diverse flora that included a myriad of ferns, tree ferns, horsetails and cordaitales. These plants formed the vast and thick coal deposits in the Gondwana Permian Basins.

The Upper Artinskian sees the appearance of several spores of probable fern origin; *Microbaculispora villosa*, *Gondisporites raniganjensis*, *Punctatisporites fungosus*, *Punctatisporites cf. Lalmatiasporites*. *Gondisporites raniganjensis* has been considered a marker spore for the Late Permian in other Gondwana Basins (Banerjee and D'Rozario 1990; Tripathi et al. 2012).

Unit IV—Upper Permian: This sub-division marks the uppermost part of the Permian system in the present study. The Striatiti are still a dominant component of the microfloral assemblage, and in many samples make up over 50% of the total pollen count. A group of 'new' spores make their appearance. These are: *Kraeuselisporites*, *Bascanisporites*, *Granulatisporites*, *Densoisporites*, *Anapiculatisporites* and *Dulhuntyispora dulhunty*. A few appear right at the top of the uppermost part of Unit IV; these are *Punctatisporites fungosus* and *Culleisporites*. Many plants do not make it beyond the Artinskian as can be seen from their individual pollen/spore distribution charts in Appendix (A.1 and A.2).

The present unit IV corresponds to Balme's (1964) *Dulhuntyispora* assemblage and to Evans' (1967) 'Stage 5'. This zone is still dominated by the Glossopteris flora and the swamp-type environment. Water levels have changed over the time span of the Upper Permian as can be seen from the overall negative correlation between the Striatiti and acritarch peaks (see pollen distribution charts for bore A and bore C). Highs in acritarchs correspond to lows in the Striatiti/Glossopterids counts. Water bodies over a large area were formed, which encouraged algae to flourish and replace the swamp vegetation of the Glossopterids for a while. A prominent peak of acritarchs is seen in the uppermost part of the Upper Permian, also corresponding to a low of Striatiti.

The alternating times of change between flooding and swamp as expressed by the pollen mark a change in the depositional environment following an episodal climatic change that brought on high rainfall resulting in flooding of the swamps, rise in water level, drowning of the swamp forests and algal dominance.

Similar results were reported by Wetering et al. (2013) from their study using $\delta^{13}C/^{12}C$ of coal lithotypes alongside floral assemblages from the Upper Permian of the Bowen Basin. Their study highlighted the interaction between wet and drying-out and their effect on the Glossopteris flora. The same study demonstrated the climatic fluctuations over time.

$\delta^{13}C/^{12}C$ study of Permian coal measures ranging in age from Sakmarian to Uppermost Permian from South India by Singh et al. (2012) and Aggarval et al. (2019) reported a correlation between C-isotope composition and extreme climatic events. The samples analysed by their study differentiated between lithologies (not *lithotypes*) in general and obtained very similar results to those reported by Wetering et al. as would be expected from same kind of vegetation. The fluctuations in isotopic composition evident in both studies has been attributed by the authors to fluctuations in the climate. The extremely isotopically 'heavy' values obtained by both studies in some of the samples is most probably the result of extensive flooding and algal dominance contributing to the organic matter (Glikson 1984). The conclusion reached by both studies demonstrated indirectly the climatic effect overriding the general vegetational influence. Strongly isotopically negative values, on the other hand, were reported by all studies of the Permian-Triassic boundary (Aggarwal et al. 2019; Vajda et al. 2020) due to the extreme atmospheric conditions following intense volcanism and meteorite impact resulting in the burial and decay of huge amounts of vegetation.

Disaccate non-taeniate pollen species seem to persist throughout the Permian with occasional peaks. They represent trees resembling today's *Araucaria* and *Podocarpus* which thrived and made up the upland flora of those times. Usually disaccate pollen average about 10%. It is thought that they are the ancestors of today's conifers such as *Araucaria, Agathis, Podocarpus* and others. These were plants that no longer required growth in or near water. As the climate became warmer they occupied the mountain sides and have been regarded as 'Upland Vegetation' (Chaloner 1958; Upshaw and Creath 1965). It has been argued from studies of successions of fossil floras that higher and more specialised type of vegetation was always present in uplands of a region long before they entered the lowlands as replacement for the older flora. Suggested reason was the diversity of physical environment in upland areas which promotes rapid evolution. This may be valid as long as no change in climate effects the lowlands when the contrary happens: An example for the latter are the *Nothofagus* forests of New Guinea which are dying with the general warming up.

Acritarchs: The importance of acritarchs when in larger than usual concentrations is by signalling a change in the depositional environment, which in turn is a function of climatic change. There are **several peaks of Acritarchs** present in the Artinskian; both in the Lower Artinskian (Biostratigraphic Unit II) and Upper Artinskian (Biostratigraphic Unit III) and several prominent peaks in the upper Permian (Biostratigraphic Unit IV). Two major peaks of acritarchs are clearly evident in the Lower Artinskian, biostratigraphic Unit II, and two major ones in the Upper Artinskian (Biostratigraphic Unit III) whereas smaller peaks are spread over the rest of the Permian sequence. These are markers for flooding of the depositional environment, lakes forming, drowning of the swamp vegetation and algal domination at times. The Upper Permian (Biostratigraphic Unit IV) starts with an acritarch peak, followed by a low. Acritarchs peaked again on and off, with no acritarch presence except two small peaks towards the end of the Permian, followed by their total absence.

References

Aggarwal N, Aggarwal S, Thakur B (2019) Palynofloral, palynofacies and carbon isotope of Permian coal deposits from the Godavari Valley Coalfield, South India: Insight into the age, palaeovegetation and palaeoclimate. Int J Coal Geol (in press)

Anderson JM (1977) The biostratigtraphy of the permian and triassic part 3. A review of Gondwanan Permian Palynology with particular reference to the Northern Karoo Basin, South Africa. Mem Bot Surv South Africa 41:1–67

Balme BE (1964) The palynological record of Australian pre-Tertiary floras. In: Cranwell LM (ed) Ancient pacific floras. Uni Hawaii Press, Honolulu, pp 49–80

Banerjee M, D'Rozario A (1990) Palynostratigraphic correlation of Lower Gondwana sediments in the Chuparbhita and Hura Basins, Rajmahal Hills, Eastern India. Rev Palaeobot Palynol 65:239–255

Chaloner WG (1958) The Carboniferous upland flora. Geol Mag 95:261–262

Cridland AA (1964) Amyelon in American coal balls. Palaontology 7:86–209

Farabee M, Taylor EL, Taylor TN (1990) Correlation of Permian and Triassic palynomorph assemblage from central Transantarctic Mountains, Antarctica. Rev Palaeobot Palynol 65:257–265

Glikson M (1984) Further studies of torbanites and coorongite using transmission electron microscopy and C-isotope analysis. Org Geochem 7(2):151–160

Götz AE, Hancox PJ, Lloyd A (2017) Permian climate change recorded in palynomorph assemblages of Mozambique (Moatiize Basin, eastern Tete Province). Acta Palaeobotanica 57(1):3–11

Guppy J, Milne P, Glikson M, Moore H (1973) Further development in computer assistance to pollen identification. Geol Soc Austr 4:201–206

Helby RJ (1967) Preliminary palynological study of "Kutung" sediments in central eastern New South Wales. Geol Surv New South Wales. Palyn Dept 5:1–7

Lord JH (1952) Collie mineral field. Geol Surv West Austr Bull 105(1):163

Low LG (1958) Collie mineral field. Geol Surv West Austr Bull 105(2):135

McWhae JRH, Playford PE, Lindner AW, Glenister BF, Balme BE (1958) The stratigraphy of Western Australia. J Geol Soc Austr 4:1–141

Mishra S, Jha N (2017) Early Permian (Asselian-Sakmarian) palynoflora from Chintalapudi Area, Godavari Graben, South India and its palaeoenvironmental implications. J Palaentol Soc India 62(2):23–40

Playford G (1959) Permian stratigraphy of the Woolanga Creek area. Western Australia. J Roy Soc W A 42(1):7–28

Singh PK, Singh MP, Pachiti PK, Kalpana PK, Manikayamba MS, Lakshminarayana G, Singh AK, Naik AS (2012) Petrographic characteristics and carbon isotopic composition of Permian coal: implications on depositional environment of Sattupalli coalfield, Godavari Valley, India. Int J Coal Geol 90:34–42

Tripathi A, Vijaya, Murthy S, Chakarborty B, Das DK (2012) Stratigraphic status of coal horizon in Tatapan-Ramkola coalfield, Chhattisgarh, India. J Earth Syst Sci 121(2):537–557

Upshaw CF, Creath WB (1965) Pennsylvanian mspores from a cave deposit in Devoniam limestone, Calloway County, Missouri. Micropalaeontology 11(4):431–448

Vajda V, McLoughlin S, Mays C, Frank TD, Fielding CR, Tevyaw A, Lehsen V, Boching M, Nicoll RS (2020) End Permian (252 Mya) deforestation, wildfires and flooding—Ancient biotic crisis with lesson for the present. Earth Planet Sc Lett 529:115875

Walker D, Milne P, Guppy J, Williams J (1968) The computer assisted storage and retrieval of pollen morphological data. Pollen Spores 10:251–262

Wetering Van de N, Esterle J, Baublys K (2013) Decoupling 13C response to palaeoflora cycles and climatic variation in coal: a case study from the Late Permian Bowen Basin. Queensland, Australia. Palaeogeogr, Palaeoclimatol, Palaeobot 386:165–179

Wheeler A, Götz AE (2016) Palynofacies patterns of the Highveld coal deposits (Karoo Basin, South Africa): clues to reconstruction of palaeoenvironment and palaeoclimate. Acta Palaeobotanica 56(1):3–15

Chapter 4
Spontaneous Combustion of Coal

4.1 Introduction

This is another aspect of Permian Gondwana coal that is rarely given significant attention in books about coal, although numerous short scientific papers dealt in the past and the present with the subject. Is it a phenomenon resulting from the Gondwana Permian vegetation that makes up the bulk of the coal? Does the Glossopteris flora diagenetic transformation render it to produce more bituminous substances than other vegetation? There are many questions regarding the break down and diagenesis of this specific flora dominating the Permian of Gondwana which are not known.

Spontaneous combustion is a major hazard in coal deposits both underground and above ground in stock piles. It is a potential danger during mining when followed by explosions.

Many questions are posed; such as what is meant by and actually takes place during spontaneous combustion of coal. What causes this phenomenon and why are *Permian coals of Gondwana* particularly prone?

Spontaneous combustion in coal piles, in exposed coal seams are a common occurring phenomenon and do not pose life-threatening incidences.

The crucial question that has not been precisely answered to this day is what is the source of the methane that suddenly appears in an underground coal mine and causes catastrophic explosions? This chapter will explain the causes for the generation of methane in underground mines and the explosions that follow. This knowledge has been used to construct a model predicting the susceptibility of a coal to this happening, and thus avoiding the loss of lives.

M. Glikson-Simpson, *Coal—A Window to Past Climate and Vegetation*,
https://doi.org/10.1007/978-3-030-44472-3_4

4.2 Naturally Occurring Spontaneous Combustion

(a) Below ground

Coal burning below ground has been detected for a long time and some had been burning for hundreds and thousands of years emitting smoke, detected above ground. Examples have been described from many parts of the world; such as the Burning Mountain (Mt Win-Jen) in New South Wales, Australia where a coal seam has been burning for thousands of years. It was part of a territory owned by the indigenous Wanaruah people and is tied up in their ancient past legends. The coal seam is 2 m thick and 30 m below ground and is detected by smoke coming out of the ground in the area, moving southward at a rate of about 1 m every year. The coal seam is part of Permian coals of the Sydney Basin. Many such occurrences have been reported from around the world.

The largest concentration of coal fires caused by spontaneous combustion, both in coal seams as well as coal piles occur in the Permian coals of India, first documented from the Raniganj coal Basin in 1865.

(b) Above ground coal fires caused by spontaneous combustion

Spontaneous combustion in stock-piles of coals is a constant hazard that can be and is controlled in well managed mines. There are special ways of building coal stock piles that prevents spontaneous combustion.

Spontaneous combustion is a chemical oxidation reaction whereby heat generated from the oxidation reaction is unable to dissipate by convection or conduction. The temperature rises and the oxidation reaction speeds up increasing the heat until the coal ignites. In most coal mines methane gas is present in varying quantities. Drainage of coal mines of methane and other gases is a specialised engineering feat well documented and usually planned and constructed with the mine going ahead. The methane already present in coal seams may have been sourced from the coal that is mined at some stage in its maturation history: As organic matter is buried deeper and is subjected to increase in temperature it matures. The temperature with depth varies in places, depending on the geothermal gradient of the depositional basin. The coal undergoes physical and chemical changes with depth and increase in temperature. These changes have been documented in many books on coal (e.g. Taylor et al. 1998). At a certain stage of the coal maturation process methane is released. The methane flows upwards through cleats and fractures and may end up in a coal mine where coal has not reached methane generation stage yet. An example is a mine in Huaibei Basin in Anhui province, China (Glikson et al. 2001) where methane is present in small amounts even though the coal that is being mined has not reached methane generation stage. Vitrinite reflectance of 0.8%Rm (Glikson et al. 2001) places the coal in the lower part (A) of the volatile bituminous rank. Both reflectance and fluorescence place the coal within the oil window, with very little potential of any thermogenic methane generation in the seams of the Huaibei mine. The low bitumen content, and relatively low maturation as indicated by reflectance of vitrinite as well as by fluorescence of liptinites, also indicates that this coal has not generated methane. It is evident that the low bitumen content in the Huaibei coal is the by-product of only light oil generation from liptinites and some vitrinites. The oils migrated out of the coal seams.

Any significant quantities of methane encountered within the coal seams of Huaibei are likely to have been sourced from coals deeper in the basin where the coal seams attained higher maturation.

This methane is routinely drained from the mines.

4.3 Methods Used in Evaluation of Coals in Their Susceptibility to Spontaneous Combustion

The various methods and techniques used to predict the propensity of coals to spontaneous combustion have been outlined by Cotterell (1997). In the last 20 years or so a large number of papers have been published presenting variations on the methods already in use, and suggested new techniques (Kaymakci and Didari 2002; Avila et al. 2014; Saffari et al. 2019; Onifade et al. 2019). The most trusted method used in predicting the susceptibility of a coal to spontaneously combust was developed by Saxena et al. (1990). Their study of the Raniganj coals in India pointed out the possible relationship between coal rank and spontaneous combustion. Samples were taken and the coal rank was established by vitrinite reflectance. The other parameter of importance recognised by the researchers was the maceral composition of the coals. The coals were heated and the lowest temperature at which the exothermic reaction could be observed to be self-propellent was taken as a parameter for measuring the spontaneous combustion propensity of the coals. This was termed the ***crossing point*** (Fig. 4.1); namely the point where the temperature reached crosses coal rank (designated by volatile matter content and/or vitrinite reflectance). In other words, the point of ignition. However, this method is based on coal rank as a major factor.

Saxena et al. (1990) using the crossing point experiment found that samples with lowest volatile matter and highest vitrinite reflectance were seen to obtain the highest cross-points, while samples with highest volatile matter and lowest vitrinite reflectance were seen to ignite at lower temperatures. The conclusion of this study was that as coalification increased, the susceptibility of the coal to spontaneous combustion decreased. Saxena et al.'s study also concluded that vitrinite-rich coals were more susceptible to spontaneous combustion than inertinite-rich coals.

Later studies (Cloke and Lester 1994; Cotterell 1997) found that the maceral composition factor in predicting spontaneous combustion is important but not straight

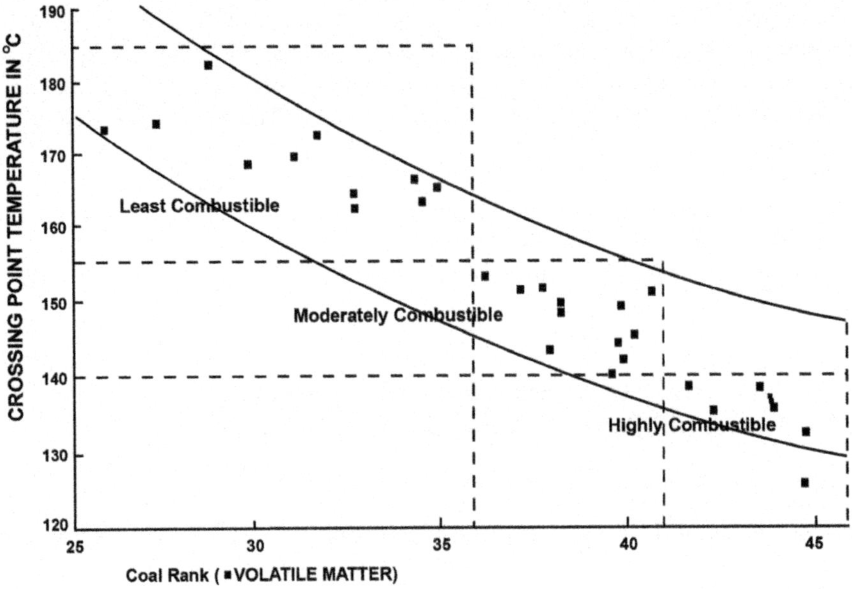

Fig. 4.1 Correlation of crossing point with volatile-matter content (after Chandra and Prasad 1990)

forward. For example, Cloke and Lester in studying spontaneous combustion of coal in power stations found that inertinite was not shown to be always inert.

Cotterell (1997) used Grewer Oven testing to determine the propensity of a Permian coal from Bowen Basin, Australia to spontaneous combustion. Cotterell determined the maceral composition of the coals by point count and the coal rank by vitrinite reflectance. The maceral composition of each sample was determined. The study found that generally the vitrinite-rich coals had the lowest auto-ignition temperature and the highest Spontaneous Combustion Index (SCI) as defined by Feng et al. (1973). However, the study found that one sample with highest inertinte content of all the coals tested had the lowest auto-ignition temperature and the highest SCI.

All studies agree that the coal rank plays an important part in the coal's propensity to spontaneous combustion. All studies are also in agreement that maceral composition plays a very important part in determining whether a coal is prone to spontaneous combustion. Maceral composition can be determined by point count technique or by density separation. However, as can be seen in Table 4.1, and Figs. 4.2, 4.3 and 4.3a by comparing the two methods, the results are not identical. This chapter will demonstrate that the precise method is the point counting one. Furthermore, the point counting method conducted by an experienced organic petrologist will recognise not just the main macerals and minerals but also the subordinate macerals, some of which play a crucial role in methane generation in a mine. Furthermore, as can be seen in

Table 4.1 Seam sample depth vitrinite liptinite inertinite minerals

	Seam	Sample	Depth	Vitrinite	Liptinite	Inertinite	Minerals
		No. 4	906	816			
Counted	No. 5a	909	895.1	42.7	14	21.3	22
Density				57.9	14.6	27.5	
Ash							28.5
Counted	No. 6	910	910.2	45.2	21.3	28	5.5
Density				45.3	23.9	30.8	
Ash							12.8
Couned	No. 7b	914	941.4	40.5	12	30.5	17
Density				52.3	11.1	36.6	
Ash							23.3
Counted	No. 10	916	1068.7	58	16.3	12.5	13.2
Density				63.5	12.1	24.2	
Ash							19.4

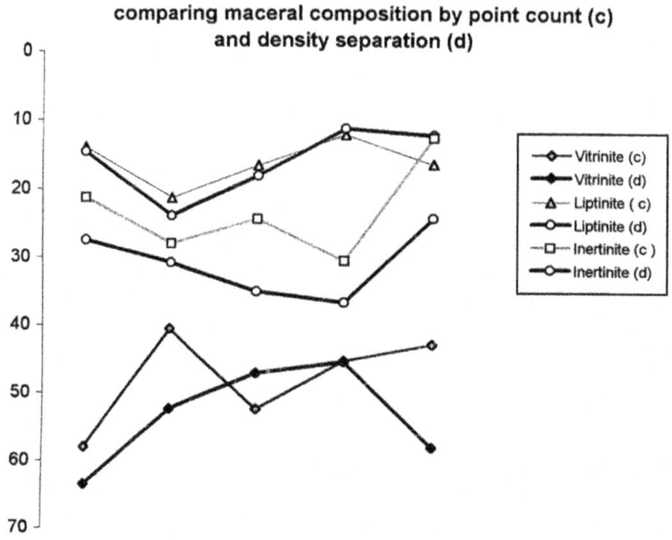

Fig. 4.2 Comparing maceral separation by two methods

the transmission electron micrographs (TEM, Fig. 4.5) minerals may be intimately associated with the organic components (maceral) and will not separate from them.

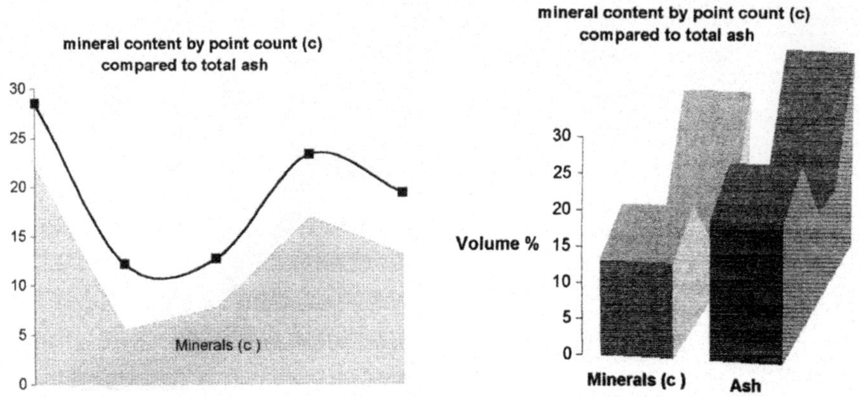

Fig. 4.3 Difference in mineral count from two methods. **a** Comparison of two methods in mineral volume

Inertinite macerals, such as fusinite for example, may contain in their cavities infill of bitumen. The bitumen is a product generated in that coal from specific macerals at a particular stage of its maturation during its burial (Taylor et al. 1998). Methane will generate from bitumen at a later maturation stage. Therefore, the presence of methane in a coal mine may have migrated from a source in a deeper buried and more mature coal. As is evident from the example of methane present in a coal having Ro of 0.8% in the Huaibei coal mine which indicated oil generated but not methane. That coal has not reached methane generation stage. However, some Permian coals in the Bowen Basin reached methane generation stage. Most coal mines are not deep enough to reach temperatures inducing methane generation from bitumen trapped in the inertinites. However, a particular coalmine in the Bowen Basin has reached 500 m, mining a coal rich in inertinite with its cavities filled with bitumen. Clearly this coal fitted perfectly into the methane generation stage. With vitrinite Ro = 1.1%, volatile matter = 30% it was ideally positioned on the coalification scale (Teichmuller 1974; Taylor et al. 1998) to generate gas upon the slightest elevation in temperature. The bitumen cracked, generated methane, spontaneous combustion occurred and ignited the methane. The result was a massive explosion. This was a tragedy of one mine.

A study by Walker et al. (2001) showed that gas in one underground Bowen Basin mine of Permian coal had a composition of 95% methane and 5% CO_2. The process of bitumen cracking into methane appears to occur at relatively low temperature (Glikson et al. 1999). The same study demonstrated experimentally that up to 95% of methane in some Bowen Basin coals is sourced from bitumens (Plate I Figs. 1–13).

COAL COMPOSITION AND MODE OF MATURATION

Plate I

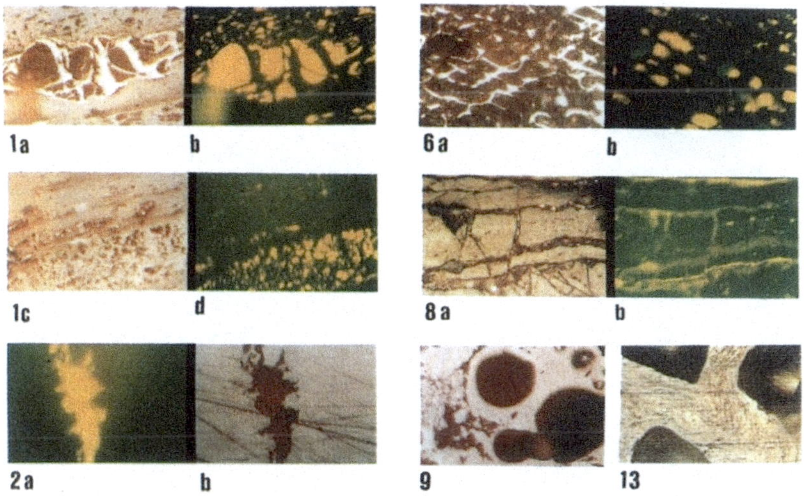

Bitumen infill of inertinite (char) cavities. Reflected white light (1a,c), and fluorescence mode (1b,d).

Bitumen infill of micro-fracture in vitrinite. Reflected white light (b), and fluorescence mode (a).

Newly generated bitumen within mosaic, formed from heated vitrinite after 600 degrees C. Reflected white light (a), and fluorescence mode (b).

Large and small pores in GOR-GCA char/inertinite formed from vitrinite after 800 degrees C. Reflected white light.

Lamellar mosaic with numerous pores characteristic of 247 coal, with initial higher reflectance (1.4%). Reflected white light.

From: Glikson et al 1999; Coalbed Methane Scientific, Environmental and Economic Evaluation Springer (Kluwer Academic Press)

4.4 Predicting the Susceptibility of a Coal to Spontaneous Combustion

A pilot study undertaken at the University of Queensland (M. Glikson and ADS Gillies) has isolated the organic components that are most prone to spontaneous combustion, as well as the dependence of dust formation on maceral composition. Previous studies (Shibaoka et al. 1985) using a hot stage microscope demonstrated

Fig. 4.4 Dark grey bands are perhydrous vitrinite bounded by cutinite (black serrated strings); light grey is standard vitrinite; white is inertinite

that during devolatilisation liptinite group macerals were the first to combust, followed by vitrinite, and inertinite in decreasing order of magnitude. Our studies have taken these experiments further and shown that the effect of macerals and secondary macerals is more complex, and may be linked to other factors.

Within the vitrinite group the susceptibility is highest in '***perhydrous*' vitrinite**' (Fig. 4.4), having a higher H/C ratio than other vitrinite group macerals, and thus closer to liptinites in chemical composition. Inertinite may also be highly susceptible to spontaneous combustion when its cavities are filled with bitumen (tar) as is not uncommon in the Permian coals. Furthermore, inertinite (char) macerals are easily pulverised into fine dust. On the other hand, vitrinite does not usually form fine dust, as it tends to break into sharp rectangular particles (Fig. 4.5).

In a pilot study of spontaneous combustion simulation, we have used whole coals, separate macerals, as well as extracted bitumen from the same coals. Bitumen showed the highest susceptibility to spontaneous combustion, and the lowest ignition temperature.

The vitrinite predominant in the Permian coals of the Bowen–Sydney Basin is a ***perhydrous*** type (Glikson and Fielding 1991; Mastalerz and Glikson 2000) equivalent to type B (Taylor et al. 1998), and to *matrix vitrinite* as defined by Crelling et al. (1988). This vitrinite may generate 'heavy' oil (bitumen) which is retained within the coal matrix.

As a result of the above-mentioned properties of the macerals, the Permian coals of the Bowen Basin contain bitumen locked within the inertinite (char) cavities (Figs. 4.6, 4.7 and 4.7a), as coatings of clay minerals and as cleat infill in vitrinite (Glikson et al. 1999). The concentrations of bitumen within the coals vary, and it

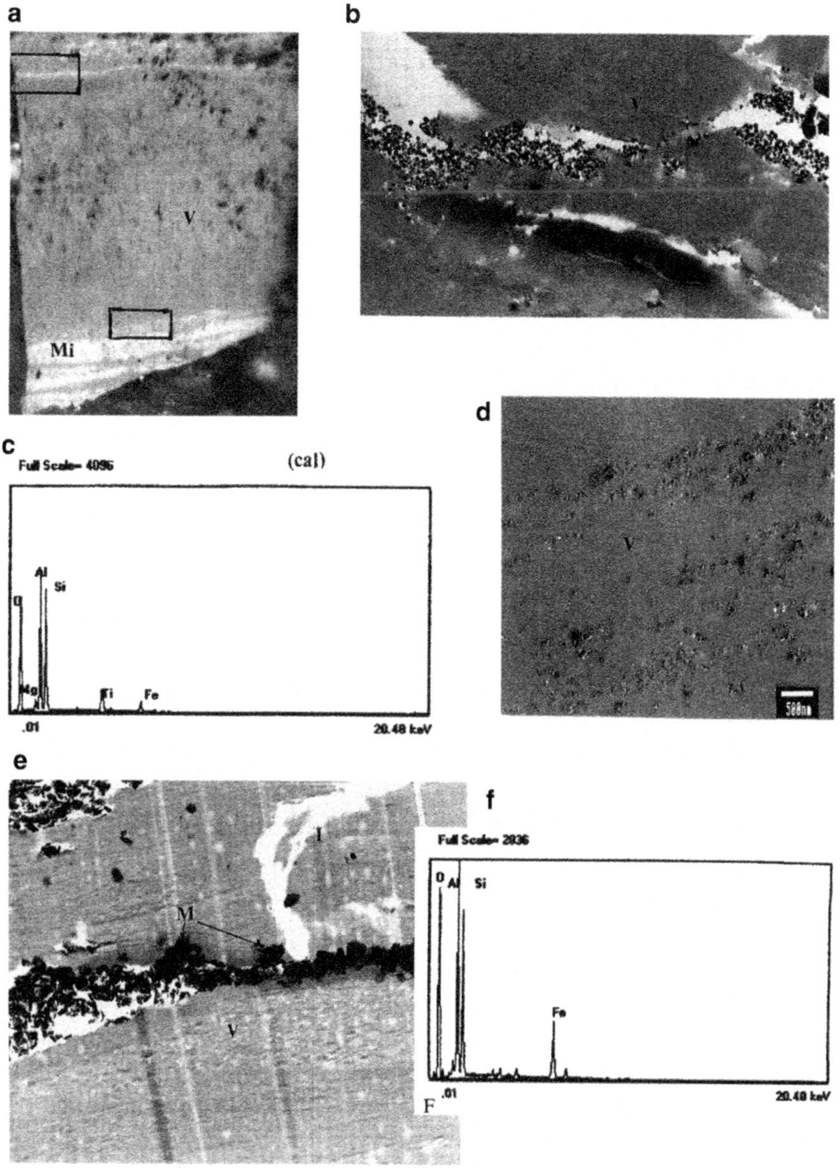

Fig. 4.5 TEM of coal macerals and closely associated minerals: (**a**) Perhydrous vitrinite (V) and micrinite (Mi). (**b**) Minerals (electron dense particles) in vitrinite. Minerals are mainly silicates as shown in EDX (**c**). (**d**) Scattered minerals throughout vitrinite. (**e**) Bands of minerals (M) in perhydrous vitrinite (V) in the form of pyrite as indicate by EDX (**e**)

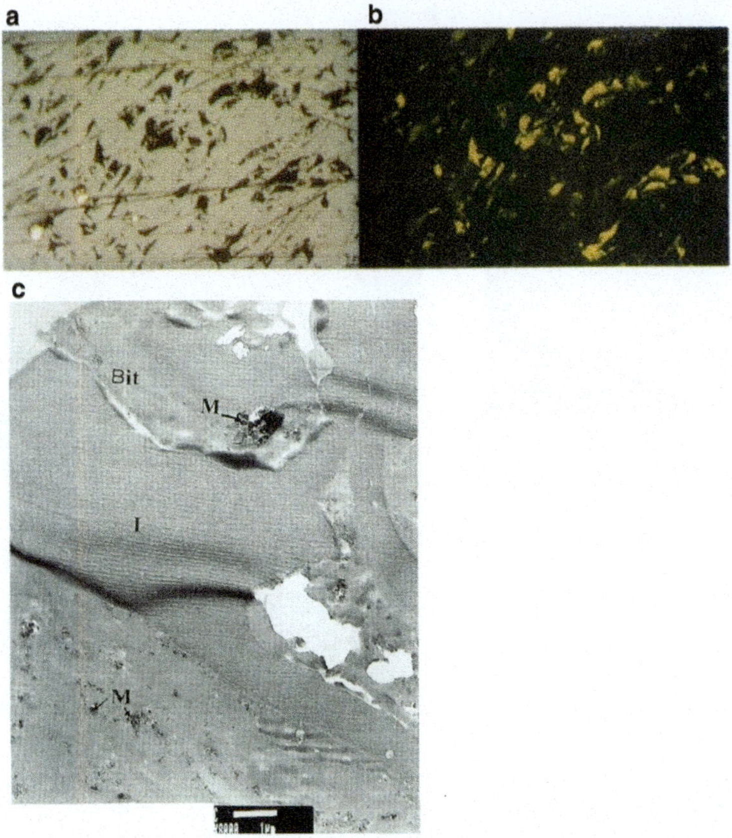

Fig. 4.6 **a** Bitumen dispersed through inertinite cavities; white light. **b** Fluorescence mode highlighting bitumen. **c** TEM : I= inertinite; M= minerals; Bit= bitumen. (M. Glikson et al., 2000)

appears from preliminary observations that some coals have higher content than others. The susceptibility of certain primary macerals and secondary macerals (e.g. bitumen/tar) to spontaneously combust, and cause of explosions appears to be also dependant on coal rank.

Concentrations of bitumen appear to be highest at a rank of 0.8–0.9 VRo (Glikson and Golding 1998). Although previous studies have shown liptinite group macerals (e.g. exinte) to be highly volatile, and therefore prone to spontaneous combustion and explosibility (Sanyal 1983), Bowen Basin Permian coals rarely contain more than 1–2% liptinite macerals. Furthermore, our pilot study showed that a Jurassic coal of sub-bituminous rank with >20% liptinite macerals (exinite and cutinite) posed no danger to explosibility due to liptinite macerals acting as a cohesive material for the coal, thus preventing any dust formation.

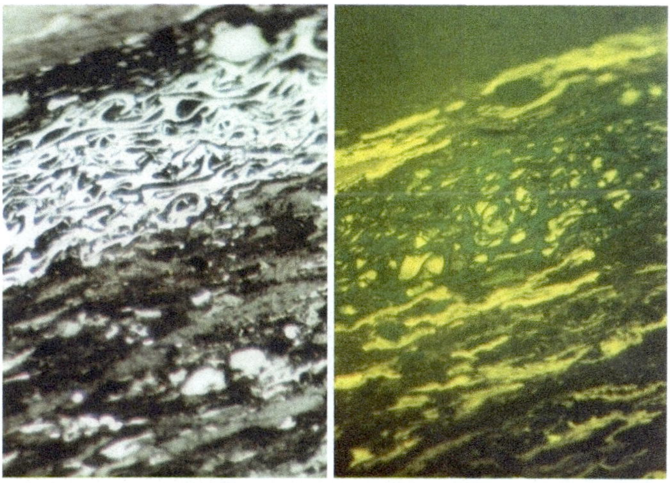

Fig. 4.7 Left; White cells are inertinite (fusinite) in reflected white light. Black areas are bitumen; narrow worm-like bands are exinite. Right; as above in fluorescence mode highlighting bitumen (inside inertinite cavities) and exinite.

4.5 Methods and Techniques Employed in the Pilot Study

It was proposed to carry out a systematic study of a range of coals of different rank from the Bowen Basin. The study was to carry out experiments on whole coal samples (after quantifying contributions from the organic and inorganic components), on isolated macerals, as well as on extracted bitumen.

Coal characterisation: Maceral composition was determined using standard techniques, as outlined in Taylor et al. (1998). Bright coal with no less than 98% vitrinite is used for pure vitrinite experiments. Likewise, inertinite was accurately characterised using Shibaoka's (1985) definitions. Total bitumen was quantified by point count. Extractable bitumen was gravimetrically quantified; extraction was carried out using chloroform in a Soxtec HT2 apparatus.

Vitrinite reflectance (Ro) of all coals was measured prior to experiments.

Testing and experimental: The testing and experiments were conducted for dust explosibility using a Mike 3 Minimum Ignition Energy Apparatus and Siwek 20 litre explosion chamber (Jackson 1995).

For spontaneous combustion, tests were carried out on the Grewer Oven (Cotterell 1997). Adiabatic oven was used in earlier combustion studies (Humphreys 1979).

The pilot study used a high-volatile bituminous coal Type A on the coalification chart (Taylor et al. 1998), Ro 0.9% which gave the results summed up in Table 4.2. These results show that the bitumen is the most explosive element, as it has the highest value of explosion pressure and pressure rise.

A study conducting **in situ** analysis of solid bitumen in coal of the Bowen Basin and Illinois Basin (Mastalerz and Glikson 2000) suggested that higher aromaticity

Table 4.2 Coal tests on separated organic components

Order (1 = highest) (4 = lowest	Explosion pressure and rate of pressure rise
	–
1	Pure Bitumen (extract)
2	Vitrinite + bitumen
3	Whole coal less extractable bitumen
4	Inertinite-rich coal

of the solid bitumen infilling cells/cavities in the Bowen Basin coal is a result of generation of gas from the bitumen. The resultant residue has become more aromatic. The same study noted that vitrinite in the Bowen Basin coals did not reveal aromatic 'carbon–hydrogen stretching', but shows distinct aliphatic stretching bands. The relatively high concentration of aliphatic compounds in the Bowen Basin vitrinites must indicate a perhydrous nature as also concluded from other studies (Glikson and Fielding 1991).

References

Avila C, Wu T, Lester E (2014) Petrographic characterization of coals as a tool to detect spontaneous combustion potential. Fuel 125:173–182

Cloke M, Lester E (1994) Characterisation of coals for combustion using petrographic analysis: a review. Fuel 73:315–319

Cotterell ME (1997) The effect of coal maceral composition on spontaneous combustion. BE Thesis, Department of Mining and Materials Engineering. The University of Queensland, Brisbane, 70 pp

Crelling JC, Skorupska MN, Marsh H (1988) Reactivity of coal macerals and lithotypes. Fuel 67:782–785

Feng KK, Chakravorty RN, Cochrane TS (1973) Spontaneous combustion; a coal mining hazard: 1973. CIM Bull 75–84

Glikson M, Fielding C (1991) The Late Triassic Callide Coal Measures, Queensland, Australia: Coal petrology and depositional environment. Int J Coal Geol 17:313–332

Glikson M, Golding SD (1998) Organic matter maturation as an indicator of hydrothermal processes in sedimentary basins. In: Arehar, Hulston (eds) Water-rock interaction, Balkema, Rotterdam, pp 101–104

Glikson M, Boreham CJ, Thiede DS (1999) Coal composition, temperature and heating rates; a determining factor in hydrocarbon species generated. In: Mastarelz M, Glikson M, Golding SD (eds) Coalbed methane; scientific, environmental and economic evaluation. Kluwer Academic Press, Dordrecht, Netherlands, pp 257–269

Glikson M, Golding SD, Boreham CJ, Saxby JD (2000) Mineralization in Eastern Australian coals: a function of oil generation and primary migration. In: Glikson M, Mastalerz M (eds) Organic matter and mineralization. Kluwer Academic Publ, pp 314–326

Glikson M., Fisher R., Golding SD, Masserotto P (2001) Coalbed Methane Project in Huaibei Basin, China: organic petrology and mineralogy. In: The 53rd meeting of the international committee for coal and organic petrology, Copenhagen, Denmark, pp 59–63

Humphreys D (1979) A Study of the Propensity of Queensland Coals to Spontaneous Combustion. ME thesis (unpublished), The University of Queensland: Brisbane

Jackson S (1995) The potential of explosibility of coal dust mixtures in oxygen deficient mine atmospheres. BE Thesis, The University of Queensland, Brisbane, pp 177–178
Kaymakci E, Didari V (2002) Relations between coal properties and spontaneous combustion parameters. Turk J Eng Environ Sci 26:59–64
Mastalerz M, Glikson M (2000) In-situ analysis of solid bitumen in coal; examples from the Bowen Basin and the Illinois Basin. Int J Coal Geol 42:207–220
Onifade M, Genc B (2019) Spontaneous combustion liability of coal and coal-shale: a review of prediction methods. Int J Coal Sci Technol 6(2):151–168
Saffari A, Sereshki F, Ataei M (2019) A comprehensive study of effect of maceral content on tendency of spontaneous coal combustion occurrence. J Inst Eng (India): Ser D, 100 (1):1–13
Sanyal A (1983) The role of coal macerals in combustion. J Inst Energy 92–95
Saxena R, Navale GKB, Chandra D, Prasad YVS (1990) Spontaneous combustion of some Permian coal seams of India: an explanation based on microscopic and physic-chemical properties. In: Jain KP, Tiwari RS (eds) Proceeding symptoms of vistas Indian palaeobotany, pp 58–82
Shibaoka M, Thomas CG, Young BC (1985) The influence of rank and maceral compositions on combustion of pulverised coal. Proc. Int. Cof. Coal Sci. 665–669
Taylor GH, Teichmuller M, Davis A, Diessel CFK, Littke R, Robert P (1998) Organic petrology. Bebruder Borntraeger, Berlin, p 704
Teichmuller M (1974) Uber neue Macerale der Liptinit–Gruppe und die Entstehung von Micrinit. Fortschr Geol Rheinld u Westf 24:37–64
Walker R, Glikson M, Mastalerz M (2001) Relation between coal petrology and gas content in Upper Newlands Seam, central Queensland, Australia. Int J Coal Geol 46: 83–92

Appendix

Details of Samples Processed for this Study

Collie Basin, Bore Site B

U.W.A. Collection No.	Lithology	
224-227	Black micaceous silty shale	60764
277-289	Black fine clayey shale	60764
297-317	Coal	60766
337-352	Shales	60767
364-379	Shales	60768
379-384	Dull coal	60769
384-392	Black shale, bright and dull coal	60770
406-426	Coal, mainly dull	60771
446	Black shale + thick band of bright coal	60772
446-488	Dark grey shale + vitrite intercalations	60773
488-494	Black, fine clayey shale	60774
513-515	Soft, micaceous silty and sandy shale	60775
530-548	White greyish sandstone	60776
548-557	Shales	60777
548-577	Coal	60778
557-577	Shales	60779
589-590	Coal	60780
591-611	Shales	60781
611-620	Shales	60782
639-658	Silty micaceous shales	60783
649-650	Coal; vitrite and fusite	60784
664-684	Black shale with bright coal bands	60785
697	Black shale with vitrite fusite	60786

(continued)

M. Glikson-Simpson, *Coal—A Window to Past Climate and Vegetation*,
https://doi.org/10.1007/978-3-030-44472-3

(continued)

U.W.A. Collection No.	Lithology	
739-759	Black silty shale	60787
771-780	Clayey shale	60788
800	Bright coal	60789
801	Clayey shale	60835
801-817	Clayey shale	60790
853-855	Silty grey shale	60791
855-875	Soft grey shale	60792
878	Black shale and coal (fusite)	60793
968-970	Bright coal	60794
973-993	Black Clayey shale	60795
1013-1033	Vitrite and fusite	60796
1040	Vitrite and fusite	60797
1078	Silty shale	60798
1079	Silty shale	60799
1112-1120	Black clayey shale	60800
1112-1120	Bright and dull coal bands	60801
1137	Black clayey shale	60802
1139	Bright coal	60803
1153-1154	Shale	60804
1229	Shale	60805
1247	Bright coal	60806
1247	Dull coal	60807
1249-1269	Clayey shale	60808
1270	Light grey sandy shale	60809
1287	Coal; vitrite and fusite	60810
1307	Vitrite	60811
1307	Durite + fusite	60812
1319	Shale	60813
1335	Vitrite + fusite	60814
1340-1361	Shale with vitrite bands	60815
1370-1380	Shale	60816
1385	Vitrite and fusite	60817
1422-1428	Dull coal; durite + fusite	60818
1468	Shale and coal interlayered	60819
1462-1463	Black shale	60820
1478-1482	Fine clayey shale	60821
1495	Shale with vitrite intercalated	60822

(continued)

(continued)

U.W.A. Collection No.	Lithology	
1505	Vitrite	60823
1509-1527	Black shale with interlayered sandstone	60824
1527-1557	Dull coal	60825
1536	Black clayey shale with thin bands of vitrite	60826
1589	Bright coal/vitrite	60827
1593	Dull coal/durite	60828
1610	Black clayey shale	60829
1630	Sandy shale	60830
1641	Shale interlayered with bright coal	60831
1670	Micaceous silty shale with carbonised plant remains	60832
1683	Dull coal	60833
1693	Shale	60834
1724	Shale	60836
1741	Shale with vitrite and fusite intercalated	60837
1815	Black shale and vitrite with carbonised plant remains	60838
1816	Grey silty shale alternating with sandstone	60839
1835	Grey silty shale alternating with sandstone	60840
1882	Dull coal with fusite	60841
1930	Bright coal	60843
1945	Black clay with vitrite and fusite interlayered	60844
1975	Shale with vitrite bands	60845
1980	Black shale	60846
2015	Shale and sandstone interlayered	60847
2047	Black shale with vitrite	60848
2059	Black shale	60849
2074	Dull coal	60850
2078	Shale with coal interlayered	60851
2083	Dull and bright coal interlayered	60852
2103	Black clayey shale	60853
2149	Black clayey shale	60854
2163	Vitrite and fusite	60855
2170	Black clayey shale	60856
2193	Bright coal/vitrite	60857
2214-2217	Black shale	60858
2242	Black shale	60859
2250	Black shale	60860
2262	Vitrite and fusite	60861

(continued)

(continued)

U.W.A. Collection No.	Lithology	
2271	Black shale	60862
2282	Black shale	60863
2294	Black shale	60864
2310	Vitrite and fusite	60865
2321	Durite	60866
2338	Micaceous silty shale	60867
2395	Black shale	60868
2418	Fine-gained sandstone with dark grey silty varves	60869
2436	Fine-gained sandstone with dark grey silty varves black shale	60870
2513	Clayey siltstone	60871
2578	Tillite	60872
2640	Tillite	60873
2060	Tillite	60874
2671	Tillite	60875
2717	Tillite	60876
2740	Tillite	60877
2770	Tillite	60878

Collie Site C

455	Shale	60888
568	Shale	60889
710	Black shale and mottled arkose	60890
1620	Shales and coal (fusite)	60891
1630	Black shale	60892
1845	Tillite with gravel	60893

Collie Site D Bore, Muja Depression

50	Coarse sandstone	60686
164-178	Coal/fusite	60687
179-18	Silty shale and black clayey shale interlayered	60689
180-201	Shales	60690
201-212	Clayey shales	60691
242-283	Clayey shales	60692
303-317	Silty shales	60693

(continued)

(continued)

50	Coarse sandstone	60686
334-345	Silty shales	60694
371-381	Black shale with layers of sandstone	60695
418-420	Black clayey shales	60696
452-462	Black clayey shales	60697
462-473	Dull coal and black shale	60698
478-489	Black shales	60699
542-552	Black shales	60700
557-577	Black shales	60701
587-597	Bright and dull coal bands	60702
618-626	Dark grey shales	60703
635-642	Coal, mainly fusite	60704
635-642	Dark grey shales	60705
695-704	Dark grey shales	60706
776-786	Dark grey shales	60707
796-806	Black clayey shales	60708
806-815	Dark grey shale with vitrite intercalations	60709
825-834	Coarse white sandstone	60710
843-846	Grey shales	60711
856-866	Grey shales	60712
875-884	Thick vitrite bands interlayered with carbonaceous shales	60713
904-913	Thick vitrite bands interlayered with carbonaceous shales	60714
944-950	Dark grey shales	60715
950-960	Black clayey shales	60716
966-976	Black clayey shales	60717
976-986	Dark grey silty shales	60718
996-1006	Dull coal	60719
1016-1034	Clayey shale	60720
1052-1069	Clayey shale	60721
1078-1090	Clayey shale	60722
1108-1125	Micaceous soft grey shale	60723
1125-1142	Kaolinitic shale	60724
1152-1166	Dull coal	60725
1176-1187	Black clayey shale	60726
1210-1216	Bright coal/vitrite	60727
1283-1301	Dull coal	60728
1307-1323	Dark grey clay with vitrite intercalations	60729
1320-1383	Black fine clayey shale	60730

(continued)

(continued)

50	Coarse sandstone	60686
1333-1351	Dull coal and clayey shale	60731/ 32
1351-1370	Shales	60733/ 34
1370-1383	Dull coal and clayey shale	60735
1370-1383	Clayey shales	60736
1397-1402	Silty sandstone	60737
1436-1455	Dull coal	60738
1436-1455	Bright coal	60739
1485-1491	Silty shale with fusite	60740
1491-1503	Clayey shale	60741
1521-1528	Black clayey shale interlayered with coal	60742
1590-1600	Coal; Dull and Bright bands	60743
1625-1631	Black shale	60744
1649-1667	Black shale with fusite	60745
1669-1684	Dull coal	60746
1695-1712	Dull coal	60747
1729-1747	Dull coal	60748
1802-1811	Coal/fusite	60749
1811-1821	Black shale	60750
1847-1865	Dull and bright coal bands	60751/52
1893-1904	Coarse white sandstone	60753
1936-1946	Dark grey clayey shale	60754
1938	White coarse sandstone	60755
1949-1967	White coarse sandstone	60756
1967-1977	White coarse sandstone	60757
2006-2016	Tillite	60758
2016-2026	Tillite	60759
2026-2036	Tillite	60760

Samples from Muja No. 1 Bore

50	Shales	60732
1126-1130	Dull coal and sandstone	60881
1135-1140	Bright coal/vitrite and shale	60879
1140-1145	Black clayey shale and bright coal/vitrite	60880

Samples from Wilga No. 3 Bore

80	Coal; vitrite and fusite	60882
110′6″	Coal; vitrite and fusite	60883
189	Black clayey shales	60884
189	Vitrite and fusite	60885
560-600	Tillite	60886
600	Tillite	60887

Samples from Alexandra Bridge Bore

1348	Siltstone	60924
1505-1515	Black shale and coal	60925
2339	Black shales	60926

Samples from Sue No. 1 Bore in Perth Basin

3991	Grey shales	60906
3998	Grey siltstone	60907
4020-4030	Greenish sandstone	60908
4093	Grey siltstone	60909
4294-4304	Greenish sandstone	60910

5497-5607	Grey siltstone intercalated with coal	60911
4901-4911	Grey sandstone	60912
5506-5511	Dark grey shales	60913
6111-6121	Grey sandstone	60914
6716-6726	Black shales	60915
6854	Black shales	60923
7595-7605	Grey sandstone with coal interlayered	60916
7897-7907	Black shales	60917
8198-8202	Coal	60918
8802-8812	Shales	60919
9102-9103	Black shales	60920
9460	Black shales	60922

Bowen Basin Samples

Samples for Part IV were collected from mining sites through the Bowen Basin, Queensland ranging in rank from 0.8 to 1.6% VR (22.5–35% VM%).

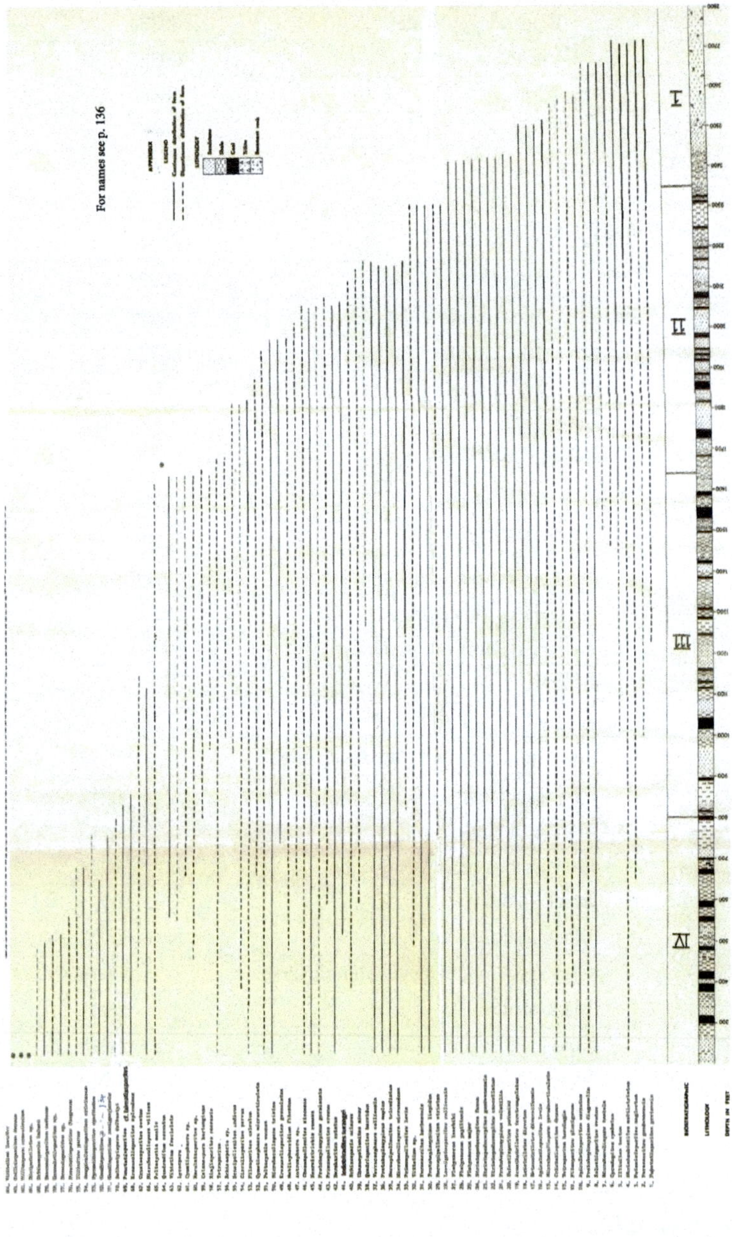

A.1 Extend of individual pollen/spores in Collie Basin, Site B. *Note:* For names see Appendix p. 136

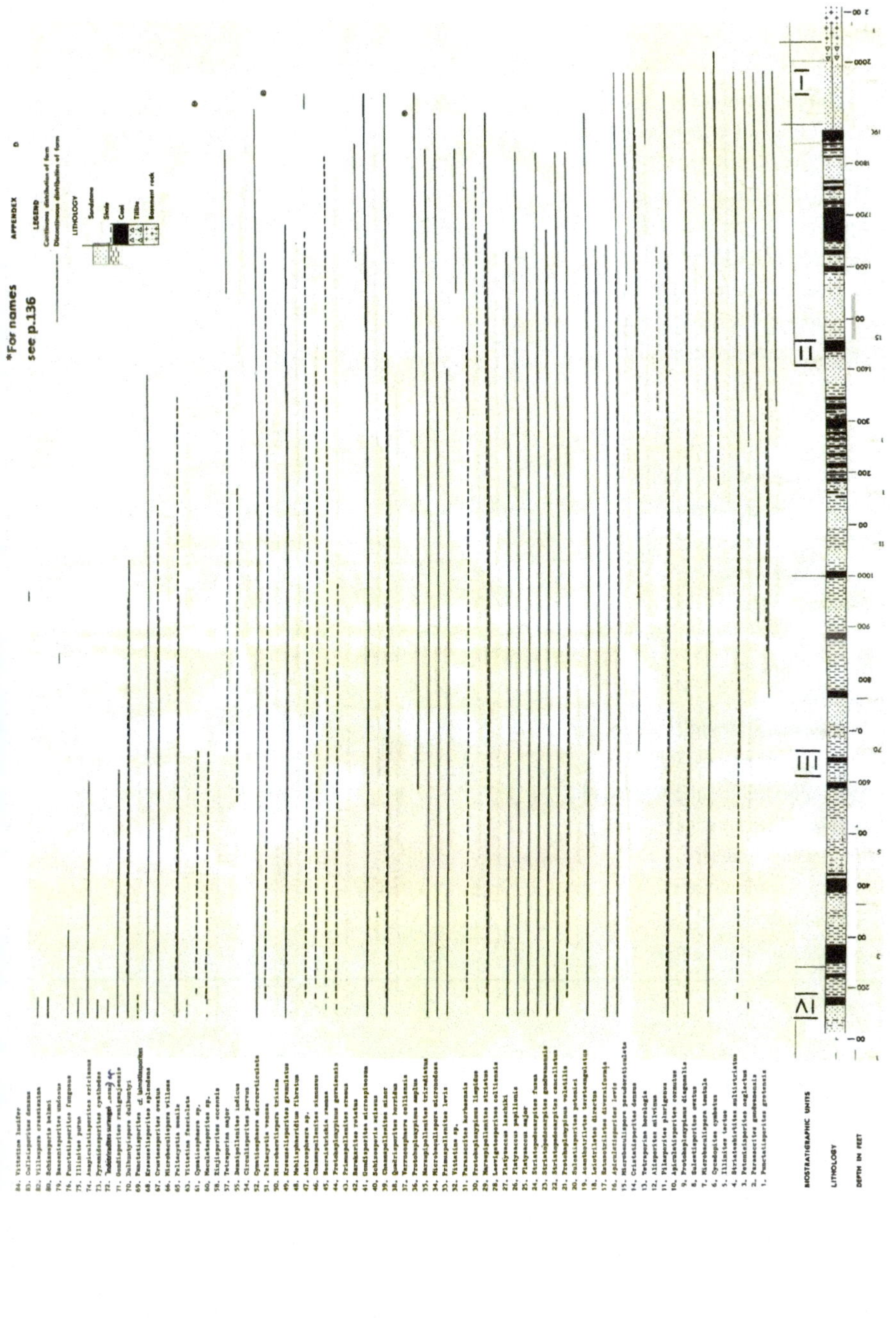

A.2 Extend of individual spores/pollen in Collie Basin, Site D. *Note:* For names see Appendix p. 136

Table 1 Key to some names of pollen/spores in Figs. A.1, A.2

1. Punctisporites gretensis	63. Vitattina
2. Parasaccites gondwanensis	66. Microbaculispora villosa
3. Potoniesporites neglectus	68. Kraeuselisporites splendens
4. Striatoabietites multistriatus	69. Puncatisporites cf almatiasporites
5. Illinites tectus	71. Gondisporites raniganjensis
6. Cycadopites cymbatus	72. Gondisporites sp.
7. Microbaculispora tentula	74. Apiculatisporites ericianus
8. Sulcatisporites ovatus	77. Densoisporites
9. Protohaploxipinus diagonalis	79. Bascanisporites undosus
10. Apiculatisporites cornutus	83. Culleisporites densus
13. Tetraporina horlogia	
15. Microbaculispora pseudoreticulata	
16. Apiculatisporites levis	
17. Retusotriletes diversiformis	
18. Leiotriletes directus	
19. Acanthotriletes tereteangulatus	
20. Sulcatisporites potoniei	
21. Protohaploxipinus volatilis	
22-24. Striatopodocarpites	
25-27. Platysaccus	
28. Laevigatosporites	
29. Marsupipollenites striatum	
30. Protohaploxypinus limpidus-amplus	
31. Parasaccites sp.	
32. Vittatina	
34. Microbaculispora micronodosa	
35. Marsupipollnites triradiatus	
36. Prtohaploxipinus amplus-limpidus	
41. Indotriradites surangei	
42. Barakarites rotatus	
45. Neoraistrickia ramose	
49. Kraeusilisporites	
50. Microbaculispora trisina	

Bibliography

Backhouse J (1993) Palynology and correlation of Permian sediments in Perth, Collie and Officer Basins, Western Australia. Rev Plaeobot Palynol 65:239–255
Balme BE (1962) Palynological reports on samples from B.M.R. 4A Canning Basin. B.M.R. Records 1962/1968
Balme BE, Playford G (1967) Late Permian plant microfossils from the Prince Charles Mountains, Antarctica. Rev Micropal 10:179–192
Bharadwaj DC (1963) Pollen grains of Ephedra and Welwotcjis and their probable fossil relatives. Mem Ind Bot Soc 4:125–135
Bahradwaj DC, Sah SCD, Tiwari RS (1965) Sporological analysis of some coal and carbonaceous shales from Barren Measure Stage (Lower Gondwana) of India. The Palaeobotanist 13:222–226
Bharadwaj, Salujha (1965) A Sporological study of Sream VII (Jote Dhemo Colliery) in the Raniganj coalfield, Bihar (India). The Palaeobotanist 13:30–41
Baharadwaj DC, Srivastava SC (1969) A Triassic microflora from India. The Palaeontographica 125:119–149
Bharadwaj DC, Venkatachala BS (1957) Microfloristic evidence on Boundary between the Carbononiferous and the Permian systems in Pfalz (W. Germany). The Palaeobotanist 6:1–11
Bolchovitina NA (1955) Paleobiology database. The fossil spore of family Gleicheniaceae. Rev Palaeobot Palynol 3(1-4):59–64
Boltenhagen E (1967) Spores et pollen du Cretace Superior du Gabon. Pollen et Spores 9(2):342–343
Butterworth MA, Williams RW (1958) The small spore floras of coal in Limestone Coal Group of the Lower Carboniferous of Scotland. Trans R Soc Edinb 63:353–392
Butterworth MA, Spinner E (1967) Lower Carboniferous spores from north-west England. Palaeontology 10(1):1–24
Chaloner WG (1953) A new species of Lepidostrobus containing unusual spores. Geol Mag 90:97–110
Chitaley SD (1951) Fossil microflora from the Mohgaon Kalan beds of the Madhya Pradesh, India. Proc Nat Ist Sci India 17:373–383
Churchill DM, Sarjeant WAS (1963) Fresh water microplankton from Flandrian (Holocene) peats of southwestern Australia. Grana Palyn 3(3):29–53
Cookson IC (1956) Pollen grains of Ephedra type in Australian Tertiary deposits. Nature 7:47–78
Crelling JC, Skorupska MN, Marsh H (1988) Reactivity of coal macerals and lithotypes. Fuel 67:782–785
Daugherty LH (1941) The Upper Triassic flora of Arizona. Carnegie Inst. of Washington Publ 526:98 pp
Deflandre G (1954) Systematique de hystrichosphaerides, sur laception de genre *Cymatiosphaera* Wetzel. C.R. Somm Soc Geol France 1954:257–258

M. Glikson-Simpson, *Coal—A Window to Past Climate and Vegetation*,
https://doi.org/10.1007/978-3-030-44472-3

Downie C, Evitt WR, Sarjeant WAS (1963) Dinofllagelates, hystrichospheres and the classification of the acritarchs. Stanford Uni Publ Geol Sci 7(3):3–16

Erdtman G (1947) Suggestions for the classification of fossil and recent pollen grains and spores. Svensk bot, Tidskr, p 41

Evans PR (1963) Spore preservation in the Bowen Basin. B.M.R. Records 100(19)

Evans PR (1969) Upper Carboniferous and Permian palynological stages and their distribution in eastern Australia. Gondwana Stratigraphy, Int Uni Geol Sci Symp. Argentina (1967) UNESCO, Earth Science (2):41–54

Florin R (1937) On the morphology of the pollen grains in some Palaeozoic pteridosperms. Svensk Bot tidsk 31(3):305–308

Glikson M, Taylor GH (1985) Cyanobacterial mats: Major contributers to the organic matter in Toolebuc Formation oil shales. Geol Soc Aust Special Publ 12:273–286

Glikson M, Lindsay K, Saxby DJ (1989) *Botryococcus*—a planktonic green alga, the source of petroleum through the ages: transmission electron microscopical studies of oil shales and petroleum source rocks. Org Geochem 14:595–608

Glikson M, Taylor D, Morris D (1992) Lower Palaeozoic and Precambrian petroleum source rocks and the coalification path of alginite. Organ Geochem 18(6):881–897

Glikson M, Duck LJ, Golding SD, Hofmann A, Bolhar R, Webb R, Baiano JCF, Sly LI (2008) Microbial remains in some earliest Earth rocks: Comparison with a potential modern analogue. Precamb Res 164:187–200

Jansonius J (1971) Emended diagnosis of Alisporites Daugherty 1941. Pollen et Spores 349–357

Klaus W (1955) Uber die Sporendiagnose des deutchen Zechsteinsalzes und des Alpinen Salzgebirgen. Z Dtch Geol Ges 105:776–788

Klaus W (1961) Some Lower Mesophytic microspores of Europe with remarks on their relation to the Gondwana-microflora. J Paleont Soc India 3(151–155)

Lele KM (1964) Studies in the Talchir flora of India. Resolution of the spore genus Nuskoisporites Potonie &. Klaus. The Palaeobotanist 12(2):147–168

Lindström S, McLaughlin S (2007) Permian event in the Prince Charles Mountains, Antarctica. Rev Palaeobot Palynol 145(1):89–122

Luber AA, Waltz (1941) Atlas of microspores and pollen grains of the Palaeozoic of the U.S.S.R. Trans. All-Union Sci Res Inst Geol 139. In: Krystofowich (ed) (1962)

Maithy PK (1965) Studies in the Glossopteris flora of India. 27 Sporae dispersae from Karbarbari beds in the Giridih Coalfield, Bihar. The Palaeobotanist 13(3):291–307

Navale GKB (1966) Petrographic and sporological studies of some coals from Talchir Coalfields, Orissa, India. The Palaeobotanost 14(2):61–68

Nilsson T (1958) Uber des vorkommen eines mesozoischen sapro-pelgesteinge in Schonene. Publ Inst Miner Paleont Quat Geol Uni Lund 53:1–111

Onifade M, Genc B (2019) Spontaneous combustion liability of coal and coal-shale: a review of prediction methods. Int J Coal Sci Technol 6(2):151–168

Pant DD (1954) Suggestions for the classification and nomenclature of fossil spores and pollen grains. Bot Rev 20:33–60

Pant DD, Srivastava GK (1965) Some lower Gondwana miospores from Brazil. Micropalaeontology 2(4):468–478

Playford G, Dettmann M (1965) Rhaeto-Liassic plant microfossils from the Leigh Creek Coal measures, South Australia. Senck Leth 46(2/3):127–181

Radforth NSW (1939) Further contribution to our knowledge of the fossil Schizaeaceae; genus senftenbergia. Trans R Soc Edinb 59:745–761

Salujha SK (1965) Miospore assemblages of seam IX of east Raniganj coalfield (India). The Palaeobotanist 13(3):227–238

Sarjeant AS (1961) Microplankton from the Kellaways rock and Oxford clay of Yorkshire. Pleontology 4(1):90–118

Schopf JW (1949) Pteridosperm male fructification: American species of *Dolerotheca*. J Paleontol 22:681–724

Schopf JW (1962) Preliminary report on the plant remains and coal of the sedimentary section in the central range of the Horlick Mountains, Antarctica. Ogio State Univ. Research Found. Inst. for polar studies, report, vol 2, 61 pp
Scott RA (1960) Pollen of ephedra from Chinie Formation (Upper Triassic) and the genus *Equisetosporites*. Micropalentology 6:271–276
Smith AHV, Butterworth MA (1954) Genus *Verrucosisporites* (Ibrahim) emended. Str Geol du Carbon Congr CR 1964:1071–1076
Srivastava SC (1970) Microfloral investigations in some coals of Talchir coalfields, Orissa, India. Palaeobotanst 18:154–166
Srivastava SC (1973) Talchir mioflora from Korba Coalfield, Madhya Pradesh. Geophytology 3:102–105
Staplin FL (1961) Reef controlled distribution of Devonian microplankton in Alberta, Canada. Paleontology 4(3):392–424
Staplin FL, Jansonius J (1964) Elucidation of some Palaeozoic densospores. Palaeontographica 114B:95–117
Staplin FL, Pocock SI, Jansonius I (1967) Relationships among gymnospermous pollen. Rev Palaeobot Palynolyn 3(1–4):297–300
Townrow JA (1960) The Peltaspermaceae, a Pteridosperm family of Permian and Triassic age. Paleontology 3(3):333–361
Venkatachala BS, Wilson LR (1963) A morphologic study and emendation of Vesicaspora Schemel 1951. Okl Geol Notes 23(6):142–149
Venkatachala BS, Kar RK (1965) Two trilete spore genera from the Permian of India. The Palaeobotanist 13(3):337–340
Walker R, Glikson M, Mastalerz M (2001) Relations between coal petrology and gas content in Upper Newlands Seam, central Queensland, Australia. Int J Coal Geol 46:83–92
Wilson LR (1959) Geological history of the Gnetales. Okl Geol Notes 19(2):35–40
Wilson LR, Venkatachala BS (1964) Morphological study of Potonieisporites Bharadwaj. Okl Geol Notes 24(3):67–68
Woodhouse RP, Wilson LR, Webster W (1946) Plant microfossils from a Fort Union coal of Montana. Am J Bot 33:271–278
Zhuo F (2017) Late Palaeozoic plants. Curr Biol 27:R853-R909, 4 pp

Index

M. Glikson-Simpson, *Coal—A Window to Past Climate and Vegetation*,
https://doi.org/10.1007/978-3-030-44472-3

The manufacturer's authorised representative in the EU is Springer
Nature Customer Service Centre GmbH, Europaplatz 3, 69115 Heidelberg,
Germany. If you have any concerns regarding our products, please
contact ProductSafety@springernature.com

Printed and bound by CPI Group (UK) Ltd, Croydon, CR0 4YY
30/09/2026
02225267-0002